AUTHENTIC OPPORTUNITIES FOR WRITING ABOUT MATH
in Upper Elementary

Teach students to write about math so they can improve their conceptual understanding in authentic ways. This resource offers hands-on strategies you can use to help students in grades 3–5 discuss and articulate mathematical ideas, use correct vocabulary, and compose mathematical arguments.

Part One discusses the importance of emphasizing language to make students' thinking visible and to sharpen communication skills, while attending to precision. Part Two provides a plethora of writing prompts and activities: Visual Prompts; Compare and Contrast; The Answer Is; Topical Questions; Writing About; Journal Prompts; Poetry/Prose; Cubing and Think Dots; RAFT; Question Quilts; and Always, Sometimes, Never. Each activity is accompanied by a clear overview plus a variety of examples. Part Three offers a crosswalk of writing strategies and math topics to help you plan, as well as a sample anchor task and lesson plan to demonstrate how the strategies can be integrated.

Throughout each section, you'll also find Blackline Masters that can be downloaded for classroom use. With this book's engaging, standards-based activities, you'll have your upper elementary students communicating like fluent mathematicians in no time!

Tammy L. Jones has taught students from first grade through college. Currently, she is consulting with individual school districts in training mathematics teachers on effective techniques for being successful in the mathematics classroom, supporting mathematics instruction, and STEM integrations. She is co-author of two book series published with Routledge: *Strategies for Common Core Mathematics* and *Strategic Journeys for Building Logical Reasoning*.

Leslie A. Texas has over 20 years of experience working with K–12 teachers and schools across the country to enhance rigorous and relevant instruction. She believes that improving student outcomes depends on comprehensive approaches to teaching and learning. She is co-author of two book series published with Routledge: *Strategies for Common Core Mathematics* and *Strategic Journeys for Building Logical Reasoning*.

AUTHENTIC OPPORTUNITIES FOR WRITING ABOUT MATH

in Upper Elementary

Prompts and Examples for Building Understanding

Tammy L. Jones and Leslie A. Texas

Routledge
Taylor & Francis Group

NEW YORK AND LONDON

Designed cover image: Getty Images

First published 2025
by Routledge
605 Third Avenue, New York, NY 10158

and by Routledge
4 Park Square, Milton Park, Abingdon, Oxon, OX14 4RN

Routledge is an imprint of the Taylor & Francis Group, an informa business

ISBN: 978-1-032-44930-2 (hbk)
ISBN: 978-1-032-44784-1 (pbk)
ISBN: 978-1-003-37457-2 (ebk)

DOI: 10.4324/9781003374572

Typeset in Warnock Pro
by codeMantra

Access the Support Material: https://resourcecentre.routledge.com/ or visit https://resourcecentre.routledge.com and search for the book's ISBN, title or authors.

Special thanks to Pixaby, WordArt, and Geometer's Sketchpad for certain images used in this book.

Online Resources

Several of the resources in this book are available online as free downloads so you can print them for classroom use. To access them, find the book at the url below and search for this book's ISBN, title, or authors. Note that you will be asked to provide information from the book before you can obtain the downloads.

https://resourcecentre.routledge.com/

You can also follow this direct link: https://resourcecentre.routledge.com/books/978032447858

Contents

Contents

Meet the Authors

Collectively, Tammy and Leslie have almost 45 years of classroom experience teaching in elementary, middle, high school, and college. This has included urban, suburban, rural, and private school settings. Being active members of their professional organizations has allowed them to continually grow professionally and model lifelong learning for both their students and their peers. In their 30-plus years of combined consulting work, they have had opportunities to work with teachers and students from kindergarten through college level. This work has spanned almost all 50 states. Their work has included helping to develop standards and curriculum at the state level as well as implementing curriculum and best practice strategies at the classroom level. One of the things that sets Tammy and Leslie apart as consultants is their work with classroom teachers, modeling and offering continued support throughout the year to build capacity at the building and district levels. Tammy and Leslie co-authored the 2013 series from Eye On Education/Routledge-Taylor & Francis Group, *Strategies for Common Core Standards for Mathematics: Implementing the Standards for Mathematical Practice* (Grades K–5, 6–8, and 9–12) and the 2017 series from Routledge-Taylor & Francis Group, *Strategic Journeys for Building Logical Reasoning: Activities Across the Content Areas* (Grades K–5, 6–8, and 9–12).

An educator since 1979, **Tammy L. Jones** has worked with students from first grade through college. Currently, Tammy is consulting with individual school districts in training teachers on strategies for making content accessible to all learners. Writing integrations as well as literacy connections are foundational in everything Tammy does. Tammy also works with teachers on effective techniques for being successful in the classroom. As a classroom teacher, Tammy's goal was that all students understand and appreciate the content they were studying; that they could read it, write it, explore it, and communicate it with confidence; and that they would be able to use the content as they need to in their lives. She believes that logical reasoning, followed by a well-reasoned presentation of results, is central to the process of learning, and that this learning happens most effectively in a cooperative, student-centered classroom. Tammy believes that learning is experiential and in her current consulting work creates and shares engaging and effective educational experiences.

Leslie A. Texas has over 25 years of experience working with K–12 teachers and schools across the country to enhance rigorous and relevant instruction. She believes that improving student outcomes depends on comprehensive approaches to teaching and learning. She taught middle and high school mathematics and science and has strong content expertise in both areas. Through her advanced degree studies, she honed her skills in content and program development and student-centered instruction. Using a combination of direct instruction, modeling, and problem-solving activities rooted in practical application, Leslie helps teachers become more effective classroom leaders and peer coaches.

We would like to give a special thanks to Trevor Styer for his work to ensure the graphics used throughout the series were high quality and reproducible for classroom use.

Preface

A Note to Our Readers

Our previous two book series, *Strategies for Common Core Mathematics: Implementing the Standards for Mathematical Practice* (Grades K–5, 6–8, and 9–12) and *Strategic Journeys for Building Logical Reasoning: Activities Across the Content Areas* (Grades K–5, 6–8, 9–12), provided a set of strategies and sample tasks that teachers could implement across the curriculum to engage students at a deeper cognitive level required by the rigorous college and career ready standards.

When we took on writing this new series, we asked ourselves: What is it that teachers want and would support students in becoming better communicators of mathematics? During training with teachers on our other two series, we often were asked how teachers could get more classroom-ready materials, such as questions, writing prompts, etc., that would support their work with students on writing and reading mathematics. Therefore, we wanted to create a collection of items for educators that would be practical and versatile, easy to implement, and yield results.

For the student, we created a collection of visual prompts that provide opportunities to engage in mathematics through looking at pictures of and from the world. There is an assortment of examples supporting the academic vocabulary associated with each math topic. Also included are ready-to-use writing prompts covering a variety of topics across the grade bands. Sets of non-typical questions are provided to promote developing a deeper understanding of mathematics. Examples of various writing styles, including creative writing, meet the needs and interests of a diverse classroom.

For educators, it is important to understand students can only become comfortable (and proficient) communicating about mathematics by practicing it regularly. Today's high-stakes assessments require students to understand mathematics in context and to explain their reasoning behind strategies and solutions. There are enough strategies included to incorporate often (daily/ weekly). Using these prompts and tasks is easy once the teacher has determined the instructional goals and targeted standards for implementation. There is teacher autonomy in implementing, but the prompts and tasks are ready to be used immediately.

These are great strategies for providing a variety of ways to engage students in mathematical discourse. The materials are versatile in use as handouts, visual displays, gallery walks, electronic documents, etc. Crosswalks show examples by mathematical topic as well as by type of writing. Teachers will find strategies for authentically integrating different writing techniques in the mathematics classroom, including creative writing. A sample lesson incorporating a number of these prompts and examples is included along with unique strategies and examples for differentiation in the mathematics classroom.

PART ONE

Why Writing in Math Matters

CHAPTER 1

Purposeful Writing: Intentional Design

Communication is essential in expressing ideas clearly and effectively. Language serves as a framework for that communication. Mathematics is often said to have its own language using symbols in addition to words. Combining mathematical language with written/spoken language can often provide deeper insight into how information is being processed, connections that are being made, conclusions drawn, etc. This data is important in assessing understanding as well as moving thinking further.

This book will look at how writing can be used in the following:

1. Making student thinking visible – formative assessment
2. Building communication skills while attending to precision – construct a viable argument and critique the reasoning of others (Standard of Mathematical Practice 3) and attend to precision (Standard of Mathematical Practice 6)
3. Establishing authentic reasons for writing, not just so we can say we did write in math.

As introduced in our book series *Strategic Journeys for Building Logical Reasoning: Activities Across the Content Areas* (Jones & Texas, 2017), there are seven opportunities for writing. These served as a guideline and informed the choices made regarding the types of writing included in this series.

- **Making Meaning** – understanding the question posed and identifying given and needed information necessary to proceed
- **Showing Evidence** – using facts and/or data to support one's argument/hypothesis/work
- **Reflecting** – being metacognitive with respect to strategies and/or processes
- **Inquiry** – creating questions to drive investigation and/or research
- **Educating** – informing others in various forms/purposes – persuasive, descriptive, expository, and narrative
- **Creating Ideas** – brainstorming/free writing to begin framing ideas
- **Producing Products** – using products to convey a message depending on audience and purpose (research papers, proposals, brochures, essays, public service announcements, etc.)

Making Student Thinking Visible: Formative Assessment

For teachers to elicit evidence of student understanding and provide feedback that moves the learning forward, students must be able to make their thinking visible. Many students struggle to organize their thoughts and capture their thinking on paper. Starting with a straightforward tool such as the Think-Write-Pair-Share (Jones & Texas, 2017) allows student-specific guidance on where to begin writing. A blank piece of paper can mean "I don't know" or "I don't care." It is an important distinction, and providing tools for students to support the "I don't know" is critical in building their capacity to help themselves. In addition, this intentional emphasis on writing highlights the importance of being a good partner by bringing something to the table when coming together to discuss ideas. Below is one example of capturing this thinking.

Think, WRITE, Pair, Share

Think about…

WRITE about what questions come to mind in the area below.

PAIR with your partner and discuss what each of you wrote.

Be prepared to **SHARE** with the whole group.

Using with a Rich Task

The following tool can be used to help students organize their thoughts around a rich task. The task can be embedded so students can stay focused while engaging in the process. The "write" component gives very specific guidance to support students whether they understand the problem or not. It also provides stems for students to consider any time they are engaged in solving a problem.

Think-Write-Pair-Share	
Think Think about the problem. INSERT TASK HERE	**Write** Write by doing one of the following: If you can solve, choose a strategy, and solve. If you cannot solve… ❑ Write all facts you know about the problem ❑ Write anything you know related to the concept addressed in the problem ❑ Write questions you have about the problem
Workspace	
Pair Pair with a partner and take turns discussing your strategies and solutions. Use this space to record strategies that were different from yours.	**Share** Share various strategies and solutions with the group. Use this space to record strategies that were different from those of you and your partner.

**See Section "Problem-Solving Process" for Sample Lesson Plan using this tool.

Building Communication Skills While Attending to Precision

The following problem-solving process and graphic organizer were introduced in *Strategies for Common Core Mathematics: Implementing the Standards for Mathematical Practice* (Texas & Jones, 2013) and can be used to assist students in making sense of problems **(Standard of Mathematical Practice #1)** as well as decontextualizing and contextualizing word problems **(Standard of Mathematical Practice #2)**. The process also requires students to construct viable arguments **(Standard of Mathematical Practice #3)** as they formulate their own ideas about the meaning of the problem and make predictions about the outcome. Once a solution is obtained, students compare to the prediction to determine the reasonableness of the solution. By giving students explicit steps to unpack the problem, they begin the process with minimal to no teacher guidance and complete the initial steps. This eliminates the blank piece of paper or the famous, "I don't know" answer. Using a consistent process over time with students will assist them in becoming better problem solvers. While this process may not always "fit" every problem, it does help students develop a systematic approach to finding the "entry point" into various tasks.

The process is like the Three Reads strategy in that it asks students to read the problem more than once. The first time they read it in its entirety to understand the context of the problem. Steps #1 and #2 then ask students to reread specific sentences as they decode the text and make sense of the problem. See below for an explanation and how the graphic organizer is used to capture the process.

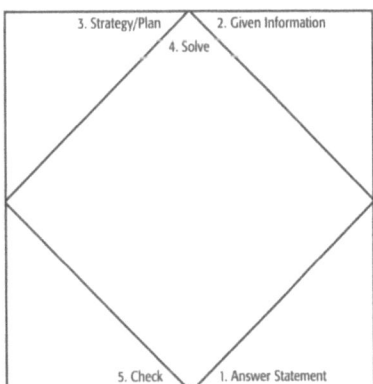

This organizer can be enlarged and copied onto paper for students. It can also be created by folding a piece of paper at each end as if making a paper airplane. Once opened, it will be partitioned as above.

For an interdisciplinary version, see our book *Strategic Journeys for Building Logical Reasoning: Activities Across the Content Areas* (Jones & Texas, 2017).

1. **Answer Statement**

   a. The question usually appears as the last sentence of the problem. Students can cover the other information and focus on the last line to determine what the problem is asking. (If the question is not here, students can check each preceding line until it is found.)

   b. Students write the question as an answer statement and leave a blank for the solution. Translating from a question to an answer statement can be challenging for some students. Practicing verbally asking and answering the question can assist in this process.

   c. Remind students to include the appropriate units for the context of the problem.

   d. The answer statement is a critical component and should be practiced even when not using this organizer. It ensures students understand the question being asked as well as guarantees they will answer the question posed if the problem contains multiple steps. Developing this habit promotes its transfer to testing situations and is particularly important when answering constructed response questions.

2. **Given Information**

   a. Students use the same process of viewing each sentence separately, covering everything else.

   b. Students determine and record relevant information from the problem.

3. **Strategy/Plan**

   a. Students use this space to state additional ideas they have about the problem, such as other information they know about the problem, possible strategies for getting started, estimations for the solution, constraints, or predictions.

   b. This is the section that allows students to formulate their own ideas about the problem and provides a place for them to create their own meaning about what is being asked.

   c. Determining an estimate also provides a context for checking for reasonableness of the solution.

   d. This step also allows students to become strategic problem solvers rather than impulsive ones by requiring them to consider the various strategies available and then determine which might be the most efficient to use in the given situation.

  e. Many students are not versatile in the various problem-solving strategies available. Creating a Strategy Wall can be useful to building the students toolkit. See p. 28 for more information on Strategy Walls.

4. **Solve**

  a. Students select a strategy (translate verbal statements into mathematical statements, draw a picture, make a table, etc.) and solve.

  b. Students can compare their solution to the estimation to determine the reasonableness of their answer.

5. **Check**

  a. Students check their answers by substitution or by using another method to justify.

  b. This is also a good time to strategically partner students who used different strategies. Students can coach each other in the use of their strategy.

  c. Once the answer has been checked, students write the answer in the blank from Step #1.

Emphasis on Process over Solution

The purpose of any problem-solving process is to encourage students to think about the problem before impulsively jumping ahead to solving. It also encourages them to read and understand before assuming what is expected. To reinforce this point, students can be given a set of problems in which they are asked to complete the initial steps but not to solve. This allows the focus to be on making sense of the problem and planning before executing. If on a teaching team, this assignment can be completed in the ELA classroom since it involves decoding text and pre-writing skills. Once students have completed these initial steps, take away the problem set and have students complete the process by solving, checking, and answering the question. By not having access to the original problems, this will serve as an assessment of the initial steps. If students can complete the work, then the information gathered was sufficient. If not, it reveals key components that were overlooked.

Update and New Information

Since the publication of *Strategies for Common Core Mathematics: Implementing the Standards for Mathematical Practice* (Texas & Jones, 2013), many teachers have asked why the graphic organizer begins at the bottom right rather than the top left. There are two reasons it is organized in this manner. The first was in response to how the brain works when asked to attend. To focus and not just mindlessly record answers in a familiar sequence/order, the brain must consciously engage with the organizer and therefore students are more intentional with the process. The second reason was addressing when the process was internalized and the tool no longer needed. Most mathematics problems begin to be solved at the top left of the problem and then worked down to the bottom right where the solution usually is completed. This organizer begins with the end in mind (bottom right) and then comes full circle with the final answer.

The problem-solving process and graphic organizer can be adapted to meet the needs of teachers and students and even eliminated as an organizer for students who internalize the process and no longer need the scaffold. Below is an example of a graphic organizer that was modified from the original. The first table contains scaffolds where there is a list of possible concepts/strategies for students to select as they build their toolkit. NOTE: The choices given here are general for illustration purposes and would be intentionally crafted for the specific unit in which it was being utilized. The second table has the supports removed.

Problem-Solving Process (Scaffolded)

The problem is asking me to... Answer Statement:	I know...
Topic/Concept this is related to... Ex. Place Value Performing Mathematical Operations Area/Perimeter Fractions Other...	Strategy for solving... Draw A Picture Guess And Check Work Backwards Use the Standard Algorithm ETC...
Solve (Show work here)	
This solution means...	

Problem-Solving Process

The problem is asking me to... **Answer Statement:**	I know...
Topic/Concept this is related to...	**Strategy for solving...**

Solve (Show work here)
This solution means...

Questioning: A Tool for Promoting Communication

As discussed in Section "Problem-Solving Process (Scaffolded)" of our second series, "Strategic Journeys for Building Logical Reasoning," there are opportunities for questioning students while working through the problem-solving process.

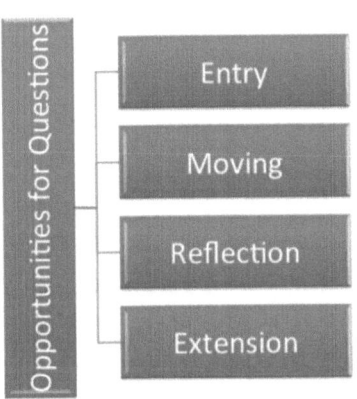

Entry: questions for students having difficulty getting started (Steps 1 and 2)
Moving: questions for places where students could get stuck (Steps 2–4)
Reflection: questions for students to use for metacognition after completing the problem/issue (Steps 4 and 5)
Extension: questions for students to engage in higher-order thinking skills with respect to the same concept and/or problem (after completing Step 5 and returning to Step 1)

These opportunities allow teachers to develop task-specific questions that can be used to support students as they are working through the process. In using these with students, it was noted that the first two opportunities occurred while students were in the middle of the process and the last two were once the process had been completed. Therefore, rather than viewing as four opportunities, they were condensed into two – I'm stuck, and I'm done.

I'm Done I'm Stuck

Four Question Types

1. **Entry Questions**

2. **Moving Questions**

3. **Reflection Questions**

4. **Extension Questions**

Task-specific questions can be generated and provided to students as needed or they can be taught to access them on their own. If stuck, they can retrieve the appropriate questions that will allow them to move forward. For students who finish early, the done questions will be used to have them go deeper with the task rather than be assigned additional work, which is oftentimes seen as busy work.

See Section "Problem-Solving Process" for examples.

Establishing Authentic Reasons for Writing

Incorporating literacy across the curriculum has long been an emphasis in mathematical classrooms. Initially, this involved having students put into words how they solved the problem alongside the mathematical steps. Fortunately, the redundancy of this request was soon realized. The mathematics itself clearly articulated what students did to solve the problem. Therefore, students were asked to write about their thinking rather than what they did. For example, if solving an equation, students were asked to write about the properties of equality used to explain why they did rather than what they did.

Section "Building Communication Skills While Attending to Precision" introduces ten different strategies that can be used to provide authentic opportunities for students to write about math. Explanations of the strategies as well as content specific examples have been provided to make these ready-to-use in the classroom. In addition, each provides the opportunity for differentiation. Below is a brief description of each:

Visual Prompts: pictures and images to initiate thoughts and discussions

Compare and Contrast: academic vocabulary word pairs to deepen understanding

The Answer Is…: giving an answer for which there could be multiple questions posed

Topical Questions: set of questions whose stems promote mathematical discourse

Writing About: using word clouds with academic vocabulary to write about a specific topic

Journal Prompts: assortment of ideas to engage students in journaling about mathematics

Poetry/Prose: collection of ideas to engage linguistic learners in expressing mathematical thought

Cubing and Think Dots: activities for independent learning

RAFT: creative writing opportunity

Question Quilts: an alternative way to present questions and provide student agency

Always, Sometimes, and Never: alternative way to view statements that promote critical thinking

PART TWO

Writing Prompts

CHAPTER 2

Visual Prompts

The visual prompts given here are actual photographs taken by the authors. These are different from what is known as "visual mathematics" which usually references the various visual representations in mathematics. The picture prompts harken back to *Mister Rogers' Neighborhood* Picture segments. These pictures help our students see the mathematics that is all around them. They also offer opportunities for students to engage in authentic mathematical communication.

The following collection of photographs can be used as journal prompts, discussion starters, bell ringers, or for centers, small groups, or learning stations. These pictures provide opportunities for students to engage in mathematics through looking at pictures of and from the world. As a starting point, have students free write what they see and describe it. This could be facilitated much like the *Notice and Wonder* prompts that the National Council of Teachers of Mathematics has brought to the forefront in the past few years.

Students can name the geometric shapes they see used, the types of numbers they see, the type of function that might mirror parts of the pictures, the mathematical topic that the image might conjure up, such as the bear outside the Denver Convention Center, and scale and proportional reasoning.

Having students free write about the visual prompts is ideal. You can make this a timed writing assignment where students must put writing implement to paper for a set amount of time, say 70 seconds. Beginning with a smaller

DOI: 10.4324/9781003374572-4

amount of time and increasing it over the semester or year will help students build stamina in free writing about a visual prompt. However, if some students need extra support, you can provide a prompt such one of these:

- ❏ What do you see?
- ❏ What are some geometric shapes you see?
- ❏ How do you think math was used in this picture?
- ❏ What questions does the picture make you think about?
- ❏ What mathematical vocabulary could you use to describe the picture?
- ❏ Do you see any patterns in the picture? If so, describe the pattern.
- ❏ Where might you have seen something similar to what this picture is showing?
- ❏ Estimate how many _____ you think are shown. How did you think about that?

Take your own pictures of things in your town or school. Look for open-source images on the internet. Remember that visual prompts offer all students a voice and provide an opportunity for most students to enter the conversation and make mathematical connections.*

* The original color images can be downloaded in the web resources.

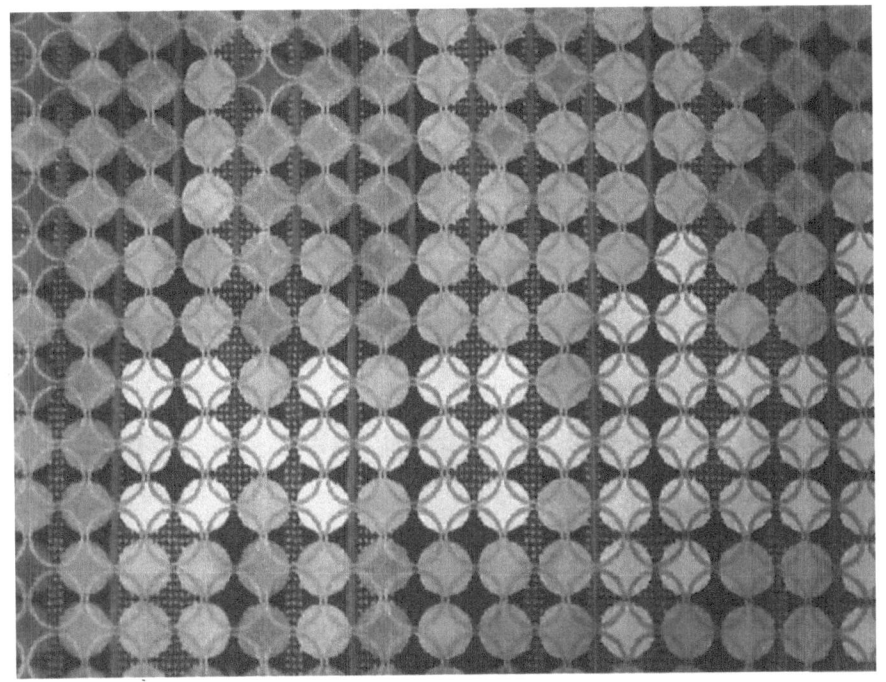

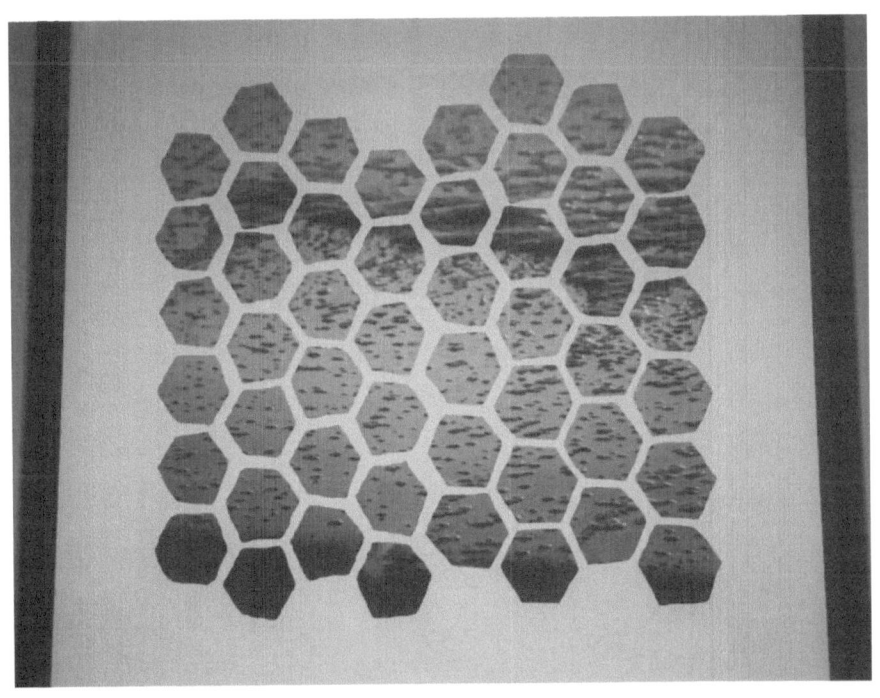

CHAPTER 3

Compare and Contrast

Writing math is typically a challenge for students. As discussed in our second series, "Strategic Journeys for Building Logical Reasoning," using the Mathematician's Notebook

> Can change the way you teach as well as how your students learn and experience their content. "The notebook becomes a dynamic place where language, data, and logical reasoning experiences operate jointly to form meaning for the student" (Jones). A Mathematician's Notebook helps students create an organized space for demonstrating their learning process. The notebook serves as a formative instructional tool as well as a portfolio of the students' learning experiences.
>
> *(Jones & Texas, p. 14)*

Whether you are using a Mathematician's Notebook, an interactive notebook, or some other method of students chronicling their journey, all students need to be writing about math daily using paper and a writing implement.

Two of the main components of the Mathematician's Notebook are the glossary and the journal. Vocabulary is one of the foundations for developing an understanding of any subject area, and mathematics is no exception. Students need many opportunities to use their vocabulary in their daily work. Having students develop a glossary and reference, the glossary as they progress through the year provides a resource for the students to use in their current

DOI: 10.4324/9781003374572-5

mathematics course as well as future courses. Additional opportunities for students to engage with their academic vocabulary are vital for students to develop the deep understanding needed for success.

One such opportunity is the Compare and Contrast activity. Students can simply make a T-chart on their paper. They write the word pair (or three columns if using three words), one word at the top of each column. Students then compare the words by listing the ways they are alike and different. They write their ideas in the columns below each word pair. They conclude by writing a summary sentence about their ideas. If time, students can complete additional pairs. There is a graphic organizer provided if desired to use. It is set up so when copied it can be cut into half and used with two students.

Interactive Word Walls and Strategy Walls

Ideally, the vocabulary used in this activity would already be displayed on a word wall of key terms that have been discussed throughout the instructional unit. A strategic way to make a word wall more interactive would be to use words from the wall for this activity. Assign students the words or allow student choice, which would reveal how students are making sense of the relationships between the concepts. Once the activity is complete, have students display their work on the wall alongside the words.

To reinforce the idea of students building a toolkit of strategies that can be used when problem solving, a strategy wall is a helpful anchor chart. Using words from the additional lists below (create a list, create a table, and create a graph, draw a picture and draw a diagram, educated guess and random guess, eliminate possibilities and solve a simpler problem, formula and function, look for a pattern and use a formula, work backwards and work forwards, write an equation or inequality and model with manipulatives), build a strategy wall at the conclusion of the activity by displaying the words (problem-solving strategies) and student responses.

A Beginning List of Word Pairs

Topic 1: Number and Quantity
Base and exponent
Equal and unequal
Equivalent fractions and simplified fractions
Expanded form and standard form
Fraction and decimal fractions
Fraction and improper fraction
Fraction and whole number

Greater than and less than
Hundreds and thousands
Like denominator and unlike denominator
Numerator and denominator
Odd and even
Ones and tens
Part of and whole
Prime and composite
Tens and hundreds
Tenths and hundredths
Unit fractions and mixed numbers

Topic 2: Algebraic Reasoning

Addition and multiplication
Addition and subtraction
Commutative property and associative property
Distributive property and associative property
Divisor and dividend
Equality and inequality
Equation and drawing
Equation and model
Expression and equation
Factors and multiples
Identity property of addition and identity property of multiplication
Identity property of multiplication and reciprocal
Multiplication and division
Product and quotient
Remainder and fractional part
Strategy and algorithm
Subtraction and division
Sum and difference
Variable and number
Zero property of multiplication and the identity property of multiplication
Zero property of multiplication and zero property of addition

Topic 3: Geometric Reasoning/Measurement and Units

2-D and 3-D
Acute angle and straight angle
Area and perimeter
Area and volume

Array and area model
Circle and sphere
Clockwise and counterclockwise
Equilateral triangle and right triangle
Horizontal and vertical
Length and width
Line segment and ray
Line symmetry and rotational symmetry
Noon and midnight
Number line and axis
Ordered pair and origin
Parenthesis and brackets
Pentagon and hexagon
Perpendicular lines and parallel lines
Polygon and regular polygon
Protractor and ruler
Quadrilateral and rhombus
Rectangle and square
Right angle and obtuse angle
Scalene triangle and isosceles triangle
Squared unit and cubic unit
Time and elapsed time
Trapezoid and parallelogram
Width and height
x-axis and y-axis
x-coordinate and y-coordinate

Topic 4: Data Analysis, Probability, and Statistics

Data and graph
Bar graph and line plot
Scaled picture graph and scaled bar graph
Interval and scale
Key and scale

Compare & Contrast

Choose a word pair. Write each word pair in the boxes below. Compare the words by listing the ways they are alike and different. Write your ideas in the columns below each word pair. Write a summary sentence about your ideas.

Word pair	
Compare:	
Contrast:	

Summary sentence(s):

Compare & Contrast

Choose a word pair. Write each word pair in the boxes below. Compare the words by listing the ways they are alike and different. Write your ideas in the columns below each word pair. Write a summary sentence about your ideas.

Word pair	
Compare:	
Contrast:	

Summary sentence(s):

CHAPTER 4

The Answer Is...

Students benefit from open-ended questions where there is possibly more than one correct response. This writing strategy allows students the opportunity to think beyond just procedural solving to get "the" answer. In some cases, the context is set up and given for the students. These questions will offer you as well as your students' insight into how they think about mathematics. Open-ended questions also encourage a growth mindset.

Students choose a card from "The Answer Is..." set to write about. Or you can assign one based upon students' individual needs. Students read the setup, if one is provided, then, they create a contextual problem for which the solution would be the answer given. This writing activity can be easily differentiated by setting parameters for students. The contextual problem can be a single step, or it may be multiple steps. It could require a specific operation or include quantities within specific parameters. Students could also be required to provide at least two different possibilities for a context where the solution is the answer. Drawings, illustrations, and labels might also be needed for a complete response.

Note that in the algebraic reasoning section, for tasks with answers such as 2½ cups, students could simply add a whole number with a basic unit fraction or double a recipe based upon their level.

Another answer shows a picture of a piece of carpet with circles that makes an array for which students need to write a possible question. The question could include the entire array, or simply a shaded section as well as whatever operation you want students to perform.

DOI: 10.4324/9781003374572-6

In the data section, data is provided from which students can generate questions as well as the answers to their questions.

Topic 1: Number and Quantity	
The answer is 1000 pumpkins. What could the question be?	The answer is 5085 ornaments. What could the question be?
The answer you get is $\frac{2}{5}$ of a candy bar. What could the question be?	The answer is $$10^7$$ What could the question be?

Topic 2: Algebraic Reasoning	
The answer is 50 legs. What could the question be?	The answer is $13\frac{3}{4}$ inches. What could the question be?
The answer you get is $2\frac{1}{2}$ cups. What could the question be?	The answer is What could the question be?

Topic 3: Geometric Reasoning/Measurement and Units	
The answer is an isosceles right triangle. What could the question be?	The answer is squares and rhombi. What could the question be?
The answer is 3 hours 45 minutes. What could the question be?	The change you receive is $47.86. What could the question be?

Topic 4: Data Analysis, Probability, and Statistics

The answer is found in the data below.

Baby Animals at the Zoo

(bar graph: y-axis "Number of Babies" 0–8, x-axis "Day of Week" Sun, Mon, Tues, Wed, Thu, Fri, Sat)

What could the question be? (Please also provide the answer(s) to your question(s).)

The answer is found in the data below.

Increase in Weight of Kenny's Puppy	
Age	Weight
Birth	5 pounds
4 months	8 pounds
8 months	13 pounds
12 months	18 pounds

What could the question be? (Please also provide the answer(s) to your question(s).)

The answer is found in the data below.

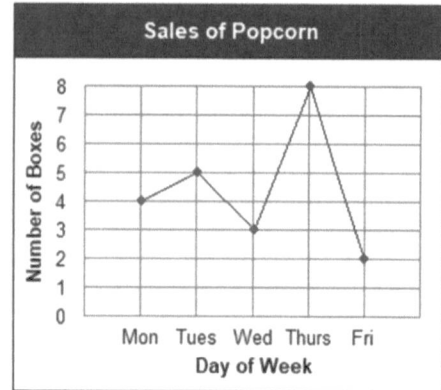

Sales of Popcorn

(line graph: y-axis "Number of Boxes" 0–8, x-axis "Day of Week" Mon, Tues, Wed, Thurs, Fri)

What could the question be? (Please also provide the answer(s) to your question(s).)

The answer is found in the data below.

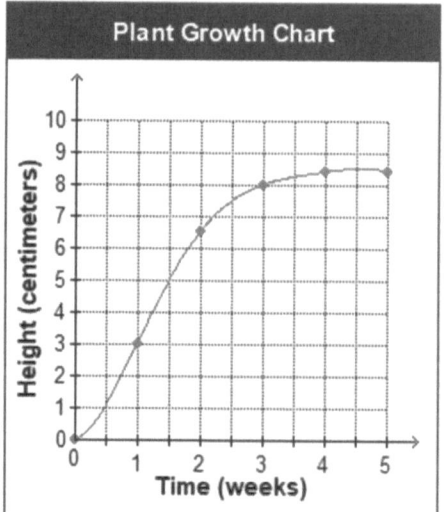

Plant Growth Chart

(line graph: y-axis "Height (centimeters)" 0–10, x-axis "Time (weeks)" 0–5)

What could the question be? (Please also provide the answer(s) to your question(s).)

CHAPTER 5

Topical Questions

Questions are tools in a teachers' toolbox and should be used as chisels to promote student thinking rather than pliers for answer-getting. Teachers practice, refine, and hone their questioning skills as they engage with students daily. Students can only provide a depth of answer based upon the quality of the question(s) asked.

As discussed in Section "Topic 3: Geometric Reasoning" of our second series, "Strategic Journeys for Building Logical Reasoning," there are opportunities for questioning students that naturally exist when students are working through a task or activity. And the questions need to be carefully crafted so the mathematical discourse is not shut down. Using a question stem such as "is," "do,", or "could" allows students the opportunity to simply answer yes or no, and then they are done. If, however, you use a stem such as "how," "what," "when," or "where" along with others listed on our Q-Pyramid overlay (Jones & Texas, p. 92), you have opened the conversation and students must engage in mathematical discourse. This use of a more inquiring form of response encourages students to justify or explain their responses, whether they be correct or incorrect.

As you hone your questioning technique, be aware that one of the most important parts of questioning is how you respond to students. You should respond to your students in a manner that supports their thinking as it reveals to you what and how they are thinking. Wait time is vital as a quick response can often shut down the individual's or rest of the class's thinking and/or reflection on what is being said. Asking students to explain why, or to further discuss

DOI: 10.4324/9781003374572-7

how they thought about something may at first be a struggle with students, but if it becomes a consistent part of your questioning, students will eventually accept it.

The questions provided in this section are not universal, but rather nuanced to the topic they reference. By design, they require communicating the answer more fully and are perfect for encouraging students to write about math. In the following, the topical examples start out with topics usually encountered in the entry level to upper elementary and extend to topics typically encountered in the upper elementary year. You can pick and choose based upon your grade level and the needs of your students.

Note: In some instances, questions may occur in more than one topic.

Topic 1: Number and Quantity

General

- ❏ How does comparing quantities describe the relationship between the quantities?
- ❏ How do numbers help us describe our world?
- ❏ What is the relationship between estimating and rounding?

Subtopic Specific

- ❏ **Place Value**
  - ❏ How does a digit's position affect its value?
  - ❏ How are place value patterns repeated in numbers?
  - ❏ How does place value affect estimation?
  - ❏ In what ways can numbers be composed and decomposed?
  - ❏ When can rounding be useful?
  - ❏ What are some situations when you should not round?
  - ❏ How are whole numbers/decimal numbers written, compared, and/or ordered?
  - ❏ How can place value patterns be used when determining a product or quotient using a power of ten?
  - ❏ How can using the relationships between numbers help solve arithmetic problems involving multi-digit numbers?
  - ❏ How can using the relationships between numbers help solve arithmetic problems involving decimals?
  - ❏ How is place value used when working with partial products?

- ❏ **Fractions**
  - ❏ How are partitioning a geometric shape and a fraction related?
  - ❏ How can fractions help us make sense of our world?
  - ❏ What is meant by a "unit fraction?"
  - ❏ What does a fraction represent?
  - ❏ What operation is represented by a fraction?
  - ❏ What is meant by equivalent fractions?
  - ❏ How are equivalent fractions identified?
  - ❏ What visual models are useful when working with fractions?
  - ❏ What are "benchmark" fractions?
  - ❏ How are benchmark fractions useful when comparing and/or ordering fractions?

- ❏ What is a "decimal fraction?"
- ❏ How can non-unit fractions be decomposed?
- ❏ How many fractions are between 0 and 1?

❏ **Exponents**

- ❏ What is an exponent and what does it represent?
- ❏ How can exponents be used to represent powers of ten?
- ❏ What is a common mistake made when operating with exponents?

Topic 2: Algebraic Reasoning

General

- ❏ Why do multiple representations of quantities help you understand the four basic operations?
- ❏ What is a pattern?
- ❏ How can features of a pattern be identified?
- ❏ How are geometric patterns and numerical patterns similar and different?
- ❏ When are grouping symbols important when working with sums, differences, products, and quotients?
- ❏ When is the "correct" answer not necessarily the best solution for the situation?
- ❏ How can using the relationships between numbers help solve arithmetic problems involving multi-digit numbers?
- ❏ How can using the relationships between numbers help solve arithmetic problems involving decimals?

Subtopic Specific

- ❏ **Operating with Numbers**
  - ❏ How are addition and multiplication related?
  - ❏ How are multiplication and division related?
  - ❏ How are division and subtraction related?
  - ❏ What are some patterns you observe in the basic multiplication and/or division facts?
  - ❏ How do those patterns help you build fact fluency?
  - ❏ How can using the properties of operations help you construct a viable argument?
  - ❏ What is the difference between factors and multiples?
  - ❏ What determines if a number is prime or composite?
  - ❏ How can tables be used to show relationships between quantities?
  - ❏ How can using the relationships between numbers help solve arithmetic problems involving two digit, three-digit numbers, etc. numbers?

- ❏ **Fractions**
  - ❏ What visual models are useful when working with fractions?
  - ❏ How can sums of unit fractions be used to represent a given fraction?

❑ How can regrouping help you determine the sums/differences of mixed numbers?

❑ How can fractions be used to solve real-world problems?

❑ How can "number sense" versus the standard algorithm help you be more efficient when operating with fractions?

❑ **Models/Equations**

❑ How are arrays used to model arithmetic operations?

❑ What are some other models that can be used to represent products and quotients?

❑ How can modeling help represent and solve problems involving multiplication and division?

❑ How can equations be used to represent contextual situations using a symbol for the unknown quantity?

❑ What does it mean to be a solution to an equation?

❑ What are some ways that operations on fractions can be modeled?

❑ How can fractions be used to represent real-world problems mathematically?

Topic 3: Geometric Reasoning

General

- ❏ How can geometry help us make sense of our world?
- ❏ What is a dimension?
- ❏ How do measuring and labeling units help us make sense of our world? Be specific.
- ❏ Why can different units represent the same measurement?
- ❏ What are some attributes of 2-dimensional shapes?

Subtopic Specific

- ❏ **Family of quadrilaterals**

  - ❏ What is a quadrilateral?
  - ❏ Where are some places quadrilaterals are used in our world?
  - ❏ How are the sides of a quadrilateral used to classify the shape?
  - ❏ How are the angles of a quadrilateral used to classify the shape?
  - ❏ How are rectangles, squares, rhombi (rhombuses) alike and different?
  - ❏ What attributes do all quadrilaterals have in common?
  - ❏ How can quadrilaterals be classified in a hierarchy based on their properties?

- ❏ **Triangles**

  - ❏ How are the sides of a triangle used to classify the shape?
  - ❏ How are the angles of a triangle used to classify the shape?
  - ❏ What attributes do all triangles have in common?

- ❏ **Angles, Lines, and Symmetry**

  - ❏ What is an angle? How are angles classified?
  - ❏ What is the relationship between a line segment, a ray, and a line?
  - ❏ How would you describe lines that are parallel?
  - ❏ What does it mean for a pair of lines to be perpendicular?
  - ❏ Can three lines be perpendicular? Explain.
  - ❏ What is a line of symmetry?
  - ❏ How can symmetry help us describe our world?
  - ❏ How are lines of symmetry determined?

❏ **Coordinate Plane**

- ❏ Who is credited with the development of the coordinate plane and for what purpose was it created?
- ❏ What is an ordered pair?
- ❏ How do you graph ordered pairs?
- ❏ How can points plotted on the coordinate plane help you identify patterns and the corresponding relationships?

Topic 4: Measurement and Units

Geometric Measurement

General

- ❏ What tool(s) is used to measure an angle?
- ❏ What unit is used when measuring an angle?
- ❏ How can patterns be used to determine standard formulas for area and perimeter?
- ❏ How are perimeter, area, and volume related?
- ❏ What is the most common way to measure an object?

Perimeter

- ❏ What does it mean to measure the perimeter of a figure?
- ❏ Discuss how two different shapes can have the same perimeter?
- ❏ How can visual models aid in the calculation of the perimeter of a figure?
- ❏ How does decomposing a complex figure into other shapes help you determine the figure's perimeter?
- ❏ How can visual models be used when calculating perimeters?
- ❏ How can the distributive property be used when calculating the perimeter of a figure?

Area

- ❏ What does it mean to measure the area of a figure?
- ❏ How does decomposing a complex figure into other shapes help you determine the figure's area?
- ❏ How can an area model be used to represent a product?
- ❏ How can an area model be used to represent the distributive property?
- ❏ How can two figures have the same area but different perimeters?
- ❏ How can two figures have the same perimeter but different areas?

Volume

- ❏ What does it mean to measure the volume of a solid?
- ❏ What is a "unit cube?"
- ❏ How can unit cubes be used to compose a solid?

❑ How can unit cubes be used to determine the volume of a solid?

❑ What are some useful tools to use when measuring liquids?

❑ How is liquid measure related to volume?

❑ **Elapsed Time**

❑ How do angles help us in telling time?

❑ How can number lines be used to represent time intervals?

❑ How can skip counting be useful when reading time?

❑ How is elapsed time measured?

❑ Where do you begin when you start to measure time?

❑ How can I model and solve problems by representing, operating with different amounts of money?

❑ How does a decimal part in an elapsed time differ from a decimal part in money?

❑ How do you convert hours to minutes? Minutes to seconds?

❑ How do the different units of time (minutes, day, weeks) relate to each other?

❑ How do you estimate the length of time of an event and know if your estimate is reasonable?

❑ **Money**

❑ How can I represent the same amount of money using different combinations of coins and bills?

❑ What are some efficient ways for combining coins and making change?

❑ Why are decimals used more than fractions when working with money?

❑ What does it mean to have "at most" $5?

❑ What does it mean to have "at least" $5?

❑ What decimal fraction would be used to represent the number of pennies you might have in terms of a dollar?

Topic 5: Data Analysis, Probability, and Statistics

General

- ❑ How can you tell is a graph contains a mistake or is intentionally misleading? (Hint: You cannot tell by just looking at the graph.)
- ❑ How does statistics help you understand the world?
- ❑ What are some ways that data is being collected on you every day?
- ❑ How can the coordinate plane be used to display data?
- ❑ When solving multi-step word problems using charts, tables, and graphs, how can you tell if the information is sufficient?
- ❑ How do you collect data?
- ❑ What kinds of questions can be answered using different data displays?
- ❑ How do you determine what data display is appropriate for a given set of data?
- ❑ How are multiple categories shown on a data display?
- ❑ How can the type of data influence the choice of the data display used?

Subtopic Specific

- ❑ **Pictographs**
  - ❑ What does "scaled" mean in terms of a pictograph?
  - ❑ How are pictographs, bar graphs, and dot plots alike?
  - ❑ What are the important parts of a pictograph? Explain each one.

- ❑ **Bar graphs**
  - ❑ Why are graphs helpful?
  - ❑ What does "scaled" mean in terms of a bar graph?
  - ❑ How are pictographs, bar graphs, and dot plots different?
  - ❑ What are the important parts of a bar graph? Explain each one.

- ❑ **Dot Plots**
  - ❑ What information does a dot plot reveal about the data set?
  - ❑ What are the important parts of a dot plot? Explain each one.
  - ❑ How does a dot plot showing data involving fractions differ from a dot plot showing data with only whole number quantities?

CHAPTER 6

Writing About...

Writing About is a small group writing activity that can be used strategically to support students who struggle with writing, particularly language learners. Just because a student can verbally tell you something does not mean that they can write that same response and support it with evidence. Prior to this activity, you might invite the ELA teacher to visit the class and share what makes a good paragraph so common expectations can be set that support the work in ELA.

Begin by giving students two or three index cards or scraps of paper. Students are to study the word cloud and write one or two sentences about the topic using words they find in the word cloud. Each student shares their sentences with the group and together create a paragraph about the topic. The index cards allow students to sequence the sentences to build a thoughtful and complete paragraph. They combine similar sentences, check for an introduction, conclusion, etc. This provides an opportunity for students to practice building a paragraph about a topic. As students first work in a group of three or four, they can then begin to work with a smaller group or a partner. The activity can be extended later as an individual writing activity as students are developing stamina for writing. Be aware that not all students will progress at the same pace.

Extensions: Students can sort the words found in the word clouds and create a mapping. Students can work in small groups, pairs, or individually. Students need to be able to articulate their sorting/mapping rule. If students

DOI: 10.4324/9781003374572-8

are doing a mapping, they can draw connectors, use string/yarn, or use something like WikkiStix™. If using WikkiStix™, be sure students are working on a piece of construction paper or scrap paper that will not matter if the sticky gets on it or not.

Once students sort their word set and show their connections, they need to write down their sorting rule in their Mathematician's Notebook. Once all groups are finished, students can do a Walk About Review where they observe the other groups' mapping/sorting and make notes about what they think their sorting rules were. Then, the whole group can come back together and discuss what they observed. Some questions that you might use to facilitate the discussion could include:

- ❏ What were the similarities you observed between the mappings?
- ❏ What were some differences?
- ❏ Were you able to identify the correct sorting rule for the other groups? Why or why not?

Suggested directions for the mapping/sort:

Study the words. Sort the words. Sort the sets of words that seem to go together. You may use your string/WikkiStix™ to show connections between the words. Explain your sorting rule fully. If directed, create a second sorting with a different rule.

Suggested directions for the Walk About Review:

As you walk about and review the other groups mappings, do not talk, look over the mapping, and in your notebook identify what you think the groups' sorting/mapping rule is and why. You will have a set amount of time at each mapping, so use it wisely and efficiently.

Ideas for Display: Groups can create a graffiti board using chart paper to capture their paragraph. These group boards can then be put together to create a graffiti wall. The class could do a gallery walk to view what was developed, provide feedback, and/or reflect on the process.

Writing about...

Study the word cloud below. Create at least two statements about data using the key words you see in the word cloud. With your group, use your sentences to create a paragraph about data.

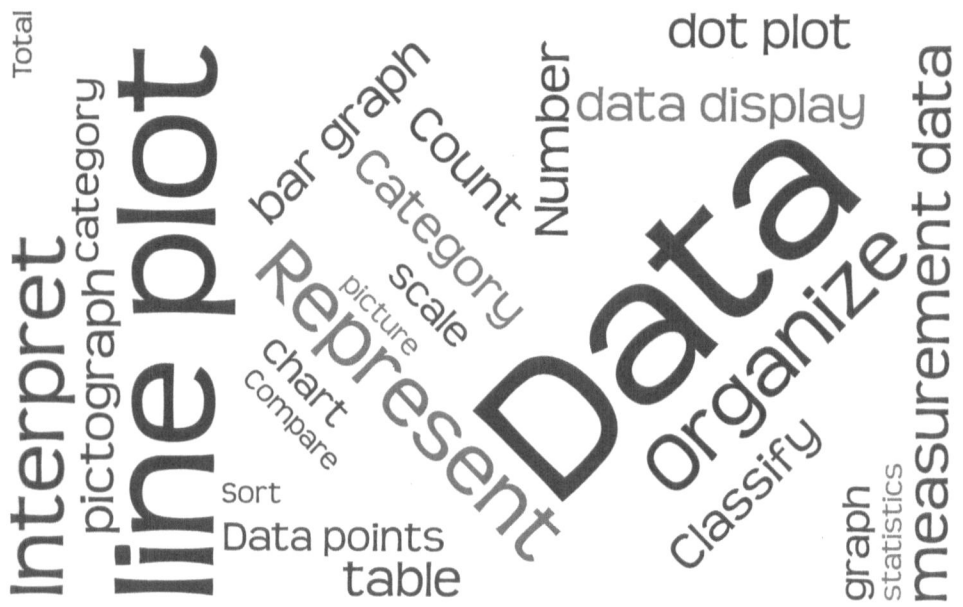

Writing about...

Study the word cloud below. Create at least two statements about fractions using the key words you see in the word cloud. With your group, use your sentences to create a paragraph about fractions.

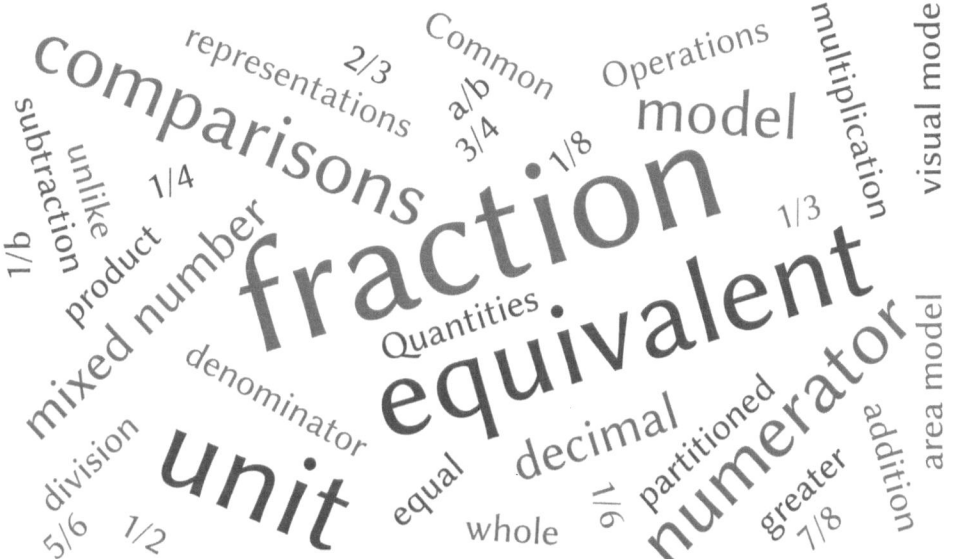

Writing about...

Study the word cloud below. Create at least two statements about angles using the key words you see in the word cloud. With your group, use your sentences to create a paragraph about angles.

Writing about...

Study the word cloud below. Create at least two statements about quadrilaterals using the key words you see in the word cloud. With your group, use your sentences to create a paragraph about quadrilaterals.

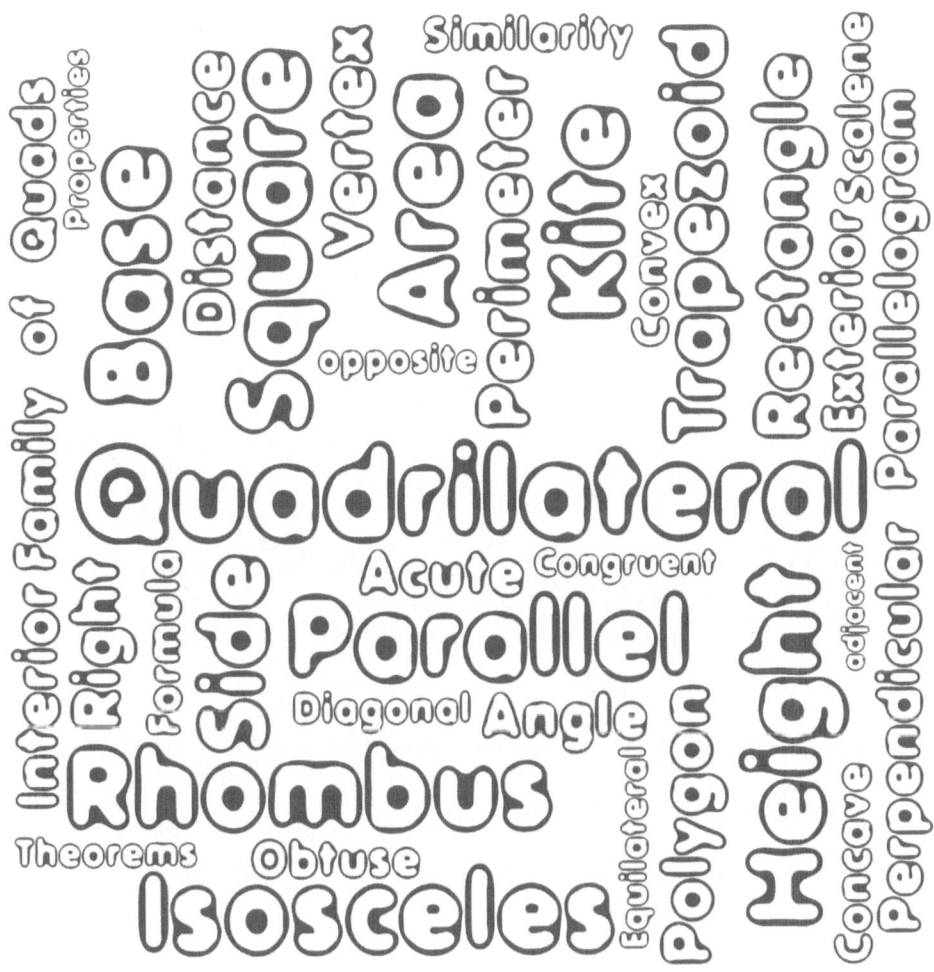

Writing about...

Study the word cloud below. Create at least two statements about triangles using the key words you see in the word cloud. With your group, use your sentences to create a paragraph about triangles.

Writing about...

Study the word cloud below. Create at least two statements about money and decimals using the key words you see in the word cloud. With your group, use your sentences to create a paragraph about money and decimals.

Writing about...

Study the word cloud below. Create at least two statements about multiplication using the key words you see in the word cloud. With your group, use your sentences to create a paragraph about multiplication.

Writing about...

Study the word cloud below. Create at least two statements about multiplication and division using the key words you see in the word cloud. With your group, use your sentences to create a paragraph about multiplication and division.

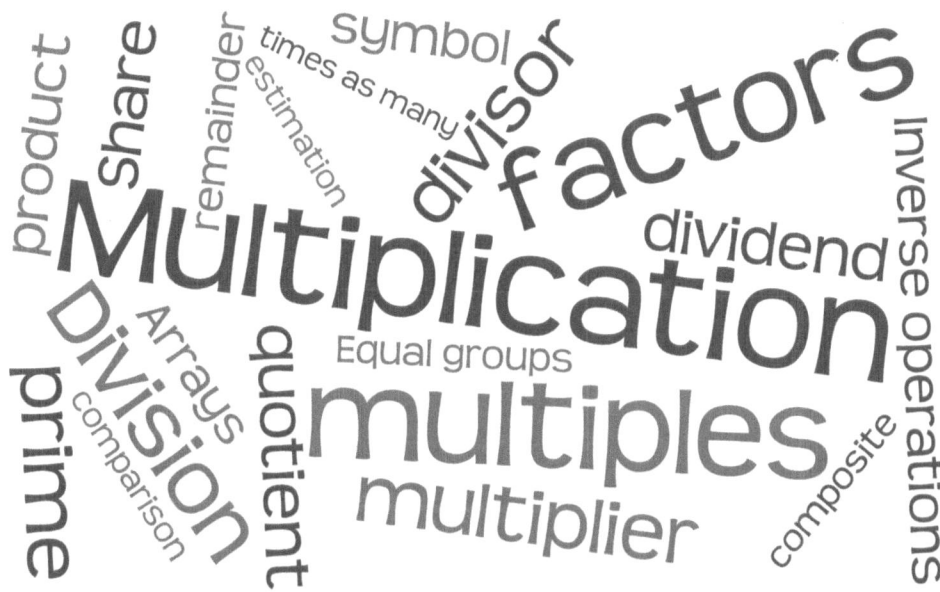

Writing about...

Study the word cloud below. Create at least two statements about place value to 100 using the key words you see in the word cloud. With your group, use your sentences to create a paragraph about place value to 100.

Writing about...

Study the word cloud below. Create at least two statements about place value using the key words you see in the word cloud. With your group, use your sentences to create a paragraph about place value.

Writing about...

Study the word cloud below. Create at least two statements about the basics of time using the key words you see in the word cloud. With your group, use your sentences to create a paragraph about the basics of time.

Writing about...

Study the word cloud below. Create at least two statements about time using the key words you see in the word cloud. With your group, use your sentences to create a paragraph about time.

Writing about...

Study the word cloud below. Create at least two statements about measurement using the key words you see in the word cloud. With your group, use your sentences to create a paragraph about measurement.

Topic 1: Number and Quantity

Area Model	**Arrays**
Base Ten	**Compare**
Compose 10	**Compose 100**

Concrete model	**Decompose**
Drawing	**Expanded form**
A Flat	**Hundred**

Hundred Thousand	**Multi-digit**
Multiples of Ten	**Number Names**
Numeral	**One-digit**

Ones	**Place Value**
Power of Ten	**Remainder**
A Rod	**Rounding**

Ten Thousand	**Tens**
Thousands	**Three-Digit**
Two-digit	**Unit**

Addition	Area Model
Common denominator	Comparisons
Decimal	Denominator

Division	**Equivalent**
Fraction	**Greater**
Mixed number	**Model**

Multiplication	Numerator
Operations	Partitioned
Product	Quantities

Representations	Subtraction
Unit Fraction	Unlike fractions
Visual Model	Whole

Topic 2: Algebraic Reasoning

Arrays	**Comparison**
Composite	**Dividend**
Division	**Divisor**

Equal groups	**Estimation**
Factors	**Inverse operations**
Multiples	**Multiplication**

Multiplier	**Prime**
Product	**Quotient**
Remainder	**Times as Many**

Topic 3: Geometric Reasoning/
Measurement and Units

Acute	**Altitude**
Angle	**Arc**
Area	**Base**

Circle	**Congruent Triangles**
Degree	**Equilateral**
End Point	**Geometric**

Interior Angles	**Isosceles**
Measure	**Obtuse**
Perimeter	**Protractor**

Rays	**Right**
Scalene	**Side**
Symbol	**Triangle**

Turns	**Vertex**
Adjacent	**Concave**
Convex	**Diagonal**

Distance	Exterior
Family of Quads	**Kite**
Parallel	**Parallelogram**

Perpendicular	Polygon
Properties	Quadrilateral
Rectangle	Rhombus

Similarity	**Square**
Theorems	**Trapezoid**
Yards	**Yard Stick**

Centimeters	Feet
Grams	Height
Inches	Indirect measurement

Kilograms	Length
Liters	Mass
Measurement	Measuring tape

Meter Stick	Meters
Non-standard	Number line
One cubic unit	Ounces

Ruler	**Standard Measurement**
Unit squared	**Volume**
Weight	**Width**

Analog	**Clock face**
Clocks	**Digital**
Half Hour	**Hands**

Hour Hand	**Hours**
Minute Hand	**Minutes**
Quarter Hours	**Elapsed Time**

$	¢
Cents	**Decimal fraction**
Decimal Point	**Decimals**

Dime	Dollar
Hundredths	**Money**
Nickel	**Penny**

Power of Ten	**Quarter**
Tenths	**Thousandths**

Topic 4: Data Analysis, Probability and Statistics

Category	**Chart**
Classify	**Compare**
Count	**Data display**

Data Points	Data
Dot Plot	Graph
Interpret	Line Plot

Measurement data	**Number**
Organize	**Pictograph**
Picture	**Represent**

Scale	Sort
Statistics	Table
Total	

Journal Prompts

As mentioned previously in the Compare and Contrast section, two of the main components of the Mathematician's Notebook are the glossary and the journal. Journals are a great way for students to keep track of their mathematical journey as well as giving you insight into how they think about math and how they have developed and grown over the course of the semester or year. Journals can be a place where students engage with quotes, historical connections to topics, famous people related to the topic of study, misconceptions, and "What if?" scenarios. If you keep a "parking lot" in your classroom for students to list issues they had with the homework from the day before, if only a couple of people had trouble with say problem #2, we might just have other students work that at the board and then see what follow-up questions, if any, were needed. But if ten students had issues with problem #2, then as the teacher, I must ask, "What did I not do that allowed students access to beginning the problem." So, that problem might go into the journal with a discussion and notes around issues students were having.

Journal entries are best assessed separately from the rest of the Mathematician's Notebook. Even a basic *All, Most, Some, None* format works well. Journals are a place that you can dialogue with students about topics and engage with them in a different way. The authors never wrote directly on the student's notebook pages, as that was their own work, but wrote on sticky notes and attached it to the page(s) where comments were appropriate.

DOI: 10.4324/9781003374572-9

Following are some beginning suggestions for journal prompts that can be used throughout the year. Some are very focused around mathematical topics, and some are just for fun for students to allow their imagination to run wild. Hopefully, these will serve as the basis for you to add many additional ideas of your own.

Math Specific

Math-ography: Students write about their prior experiences with mathematics. No comments on specific teachers allowed. What topics made sense, what topics were challenging, and how do they see mathematics fitting into their occupational plans.

WRITE YOUR "*MATH-OGRAPHY.*"

Include your:

- Earliest remembrances of counting, learning about numbers
- Elementary school work, topics you "got" as well as topics that were challenging
- Your goals for this class
- What you see yourself doing after you complete high school and what role math will play in that....

DO NOT name teacher names!

A self-evaluation: Usually assigned around the first interim period for grading, this provides the opportunity for students to reflect on their work so far. It has proved helpful to give students a list of questions to guide their reflections. Some examples are:

- ❑ Do you have a dedicated place at home to study?
- ❑ Have you been regular in your attendance?
- ❑ How would you rate your engagement in class when you are here?
- ❑ How much time outside of class are you spending on work for this class?
- ❑ Do you feel that you are getting enough sleep and rest?
- ❑ How are your eating habits affecting your schoolwork?

Writing to Explain

Option 1: Students write an explanation for a student in their class who was absent the day they learned about/how to (insert topic/activity/procedure for the day here.)

Option 2: Students area assigned a mathematical topic such as one from geometry or place value. Then they complete the following:

- ❏ You are a triangle. Tell us everything we should know about you.
- ❏ You are a fraction. Tell us everything we should know about you.
- ❏ You are the ten thousands place in a number. Tell us everything we should know about you.

Creative Writing

Students complete each of the following prompts using their imagination. Encourage students to just not write but to also use drawings and sketches and color as they complete the prompt.

- ❏ If I were a fraction, I would be _____ because....
- ❏ If I were a quadrilateral, I would be _____ because....
- ❏ If I were a mathematical pattern, I would be _____ because....
- ❏ If I were a mathematical operation, I would be _____ because....
- ❏ What I find the most challenging with _____ (current topic) is.... Explain why.
- ❏ When I see a math problem with words, I feel _____ because....
- ❏ Choose a character from a book, movie, or tv show and describe how he/she might use mathematics in what he/she does.
- ❏ Write a compliment to yourself for something you accomplished in math recently.
- ❏ Go outside and enjoy the natural world. Find two objects from nature. Keep one and give one to another person. Draw and write about the two objects you found, using math words when appropriate. Share your drawings and your thoughts.
- ❏ Write a short story where zero and $\frac{2}{5}$ are the main characters.

Using Quotes

Option 1: Students write the quote, they write what it meant in the time it was written, and then how it would be applicable to them in math class today.

Example: *Arithmetic is being able to count up to twenty without taking off your shoes. – Mickey Mouse*

- ❏ Students copy the quote in their journal.
- ❏ They then write a couple of sentences about what they think Mickey meant.
- ❏ Students conclude by writing a couple of sentences about how this will apply to their work in math class. When writing in mathematics, in their Mathematician's Notebook, etc. students need to learn to not only be precise, but to be succinct with their explanations and to the point.

Option 2: Students write a response to the author.

Option 3: Students write about what questions the quote prompts them to think about.

Option 4: The students describe what the quote means to them.

Below is a beginning list of quotes which span from historical to modern day, includes a diverse group of individuals, and cuts across disciplines to include the humanities as well as the sciences.

That awkward moment when you finish a math problem and your answer isn't even one of the choices. – Ritu Ghatourey[1]

There is a fine line between a numerator and a denominator; only a fraction will understand. – Anonymous[1]

The only way to learn mathematics is to do mathematics. – Paul Hamos[1]

Math: the only subject that counts. – Anonymous[1]

It's not that I'm so smart; it's just that I stay with problems longer. – Albert Einstein[1]

There should be no such thing as boring mathematics. – Edsger Dijkstra[1]

If you can't explain it simply, you don't understand it well enough. – Albert Einstein[1]

Mathematics is the music of reason. – James Joseph Sylvester[1]

What is mathematics? It is only a systematic effort of solving puzzles posed by nature. – Shakuntala Devi[1]

You never fail until you stop trying. – Albert Einstein[1]

I've always been interested in using mathematics to make the world work better. – Alvin E. Roth[1]

Mathematics is, in its way, the poetry of logical ideas. – Albert Einstein[2]

Mathematics knows no races or geographic boundaries; for mathematics, the cultural world is one country. – David Hilbert[2]

You don't have to be a mathematician to have a feel for numbers. – John Forbes Nash, Jr.[2]

Millions saw the apple fall, but Newton asked why – Bernard Baruch[2]

I have always enjoyed mathematics. It is the most precise and concise way of expressing an idea. – N. R. Narayan Murthy[2]

Nature is written in mathematical language. – Galileo Galilei[2]

Mathematics is not a careful march down a well-cleared highway, but a journey into a strange wilderness, where the explorers often get lost. – W.S. Anglin[2]

Sometimes the questions are complicated and the answers are simple. – Dr. Seuss[2]

Pure mathematician just love to try unsolved problems, they love a challenge. – Andrew Wiles[2]

Do not worry too much about your difficulty in mathematics, I can assure you that mine are still greater. – Albert Einstein[2]

The essence of mathematics is not to make simple things complicated, but to make complicated things simple. – Stan Gudder[2]

Mathematics consists of proving the most obvious thing in the least obvious way. – George Pólya[2]

Math is the language of the universe. So, the more equations you know, the more you can converse with the cosmos. – Neil deGrasse Tyson[2]

Number rules the universe. – Pythagoras[2]

To not know math is a severe limitation to understanding the world. – Richard P. Feynman[2]

It is impossible to be a mathematician without being a poet in soul. – Sofia Kovalevskaya[3]

The pure mathematician, like the musician, is a free creator of his world of ordered beauty. – Bertrand Russell[3]

Go down deep enough into anything and you will find mathematics. – Dean Schlicter[4]

Mathematics is the most beautiful and most powerful creation of the human spirit. – Stefan Banach[4]

The essence of mathematics lies in its freedom. – Georg Cantor[4]

A mathematician who is not also something of a poet will never be a complete mathematician. – Karl Weierstrass[4]

No human investigation can be called real science if it cannot be demonstrated mathematically." – Leonardo Da Vinci[4]

Wherever there is a number, there is beauty. – Proclus[4]

I've always enjoyed mathematics. It is the most precise and concise way of expressing an idea. – N.R. Narayana Murthy[4]

Mathematics is the science which uses easy words for hard ideas. – Edward Kasner[4]

Mathematics is the gate and key to science. – Roger Bacon[4]

But in my opinion, all things in nature occur mathematically. – Rene Descartes[4]

The important thing to remember about mathematics is not to be frightened – Richard Dawkins[4]

In mathematics, the art of proposing a question must be held of higher value than solving it. – Georg Cantor[4]

Mathematics is a place where you can do things which you can't do in the real world. Marcus du Sautoy[4]

Obvious is the most dangerous word in mathematics. – Eric Temple Bell[4]

We will always have STEM with us. Some things will drop out of the public eye and go away, but there will always be science, engineering, and technology. And there will always, always be mathematics. – Katherine Johnson[4]

You don't have to be a mathematician to have a feel for numbers. – John Forbes Nash[4]

References

1. https://brighterly.com/blog/math-quotes-for-kids/
2. https://www.playosmo.com/kids-learning/math-quotes-for-kids/
3. https://www.mashupmath.com/blog/inspirational-math-quotes
4. https://www.splashlearn.com/blog/brilliant-math-quotes-that-you-can-share-to-inspire-students/

CHAPTER 8

Poetry/Prose

In the spirit of writing in response to a quote (described in the Journal Prompts), this section begins with one from JoAnne Growney's blog Intersections – Poetry with Mathematics

> Mathematical language can heighten the imagery of a poem; mathematical structure can deepen its effect.

The precision of language required of both disciplines makes the intersection of mathematics and poetry seems almost obvious. Providing the opportunity to see the connection allows students to explore this relationship while deepening their understanding of mathematics and writing skills.

Acrostic: explain – one word, expression, describing the main word...

Example:

Circles

Centers
Inside
Radii and
Chords
Lopsided never
Elegant with perfect
Symmetry

DOI: 10.4324/9781003374572-10

Math

Mysterious to some.
Ability required.
The language of our world.
Hurts my brain!

Beginning List of Terms: (See Word Lists in Compare and Contrast for Additional Terms)

Equation
Expression
Geometry (or any specific shape or term)
Inequality
Math
Numbers (or use any specific number set)
Probability
Proportional
Ratio
Statistics

Fibonacci poem: Students can create their own "Fibonacci" poem where each line of the poem has the number of words as found in the sequence. This can be differentiated for students by having them use only the first three or four numbers found in the sequence or more if their writing skills allow. The topic of the poem can be of their choosing or can be assigned. The following poem models 1, 1, 2, 3, 5, 8.

Triangles

Pointed
Three sides
And three angles
Walking from vertex to vertex
Around the corners, either obtuse, acute, or right.

Haiku: has three lines, five syllables in first and third lines and seven syllables in the second line

Example:

Ratios

Ratios compare
Proportional reasoning
Ratios equal

Pi poem: Is similar to the Fibonacci poem. Students write lines based on the digits in pi: 3.1415926535 8979323846 2643383279. This can be differentiated for students by having them use only the first three or four digits found in pi or more if their writing skills allow.

Students can also have fun with the topic as in the example below.
Example:

Pi

My favorite pie,
cherry.
Flakey crust, tart flavor,
Yum!

Free verse/free write: There is no specific form, meter, or rhyme scheme. Students can have the freedom to write as they feel. Some suggestions are given below for types of free writes students may enjoy doing.

Cartoons

Commercial, Infographic, Public Service Announcement (PSA)
Free Verse Poem
Graphic novel, for example, The Adventures of Slope Boy
Historical Fiction
Math Carols and/or seasonal songs
Math words to a current song
Short Story

CHAPTER 9

Cubing and Think Dots

Cubing and think dots are two strategies for differentiation in the classroom. Traditionally students are given a cube with a variety of activities or tasks at a given level. Different cubes can contain different levels of tasks and activities. Think dots work in a similar way. Cards with a certain number of dots are provided as well as a number cube. Students roll the number cube and work the associated activity or task on the card with the corresponding number of dots. Again, tasks and activities are varied or leveled to meet the needs of the students. You can choose the parameters for your students or create your own set of think dot cards using index cards and practice problems from your chosen curriculum.

As students are working through the various activities, emphasize that they are just not to "show their work" but to also make statements explaining and supporting their reasoning and thinking. Students need to be able to use precise mathematical language and symbols in their written work as well as clearly articulate their thinking.

Preparation

Print the cubes and think dot cards on cardstock or heavy paper. Build the cubes carefully, using the dotted lines as the folds. Tape or glue the edges together. Fold the think dot cards and tape together. These resources can be used in a variety of ways in a center or learning station as well.

Materials List

- ❑ Set of think dot cards
- ❑ Action Cube
- ❑ Representation Cube
- ❑ Operations Cube
- ❑ Digits Decahedron to make
- ❑ Multiples of 10 Decahedron to make
- ❑ Number cubes and/or dominoes (Note: You can purchase dice for games that are in various shapes beyond the basic cube and domino sets that show numbers greater than six.)

Fractions

In this adaptation of a cubing and think dots activity, there are two cubes. One cube has representation of fractions. Another cube has actions to perform with fractions. Students also use one or two number cubes, or they can use dominoes to create the fractions. There are a variety of activities that students can engage in one set of six think dot cards. Students are using both the cubes and the cards together.

Run off the cubes and cards on cardstock. Build the cubes. Fold the think dot cards and tape, glue, or laminate. These resources can be used in a variety of ways in a center or learning station as well.

Students first roll a number cube to choose which dot card they are going to use. Then, they follow those directions. Sample options for each of the cube choices are explained below.

Action Cube Options

Add/Subtract: If students roll "Add/Subtract," and they roll a three on a number cube, they follow the directions on the dot card with the three dots. They can roll the number cube twice to generate their fractions, once for the numerator and once for the denominator. Alternatively, they can pick a domino and use one side for the numerator and one for the denominator. The students then add and/or subtract their two fractions using the representation they rolled on the Representation Cube and explain their thinking.

Multiply/Divide: If students roll "Multiply/Divide," and they roll a two on a number cube, they follow the directions on the dot card with the two dots. They can roll the number cube twice to generate their fractions, once for the numerator and once for the denominator. Alternatively, they can pick a domino and use one side for the numerator and one for the denominator. The students then multiply and/or divide their two fractions and explain their thinking.

Equivalent: If students roll "Equivalent," and they roll a one on a number cube, they follow the directions on the dot card with the one dot. They can roll the number cube twice to generate their fraction, once for the numerator and once for the denominator. Alternatively, they can pick a domino and use one side for the numerator and one for the denominator. The students then represent equivalent fractions using the representations they rolled on the Representation Cube.

Name: If students roll "Name," and they roll a four on a number cube, they follow the directions on the dot card with the four dots. They can roll the number cube twice to generate their fractions, once for the numerator and once for the denominator. Alternatively, they can pick a domino and use one side for the numerator and one for the denominator. The students then name their fraction and write a letter to a peer who has been absent explain how they arrived at the name and what the name represents.

Compare: If students roll "Compare," and they roll a five on a number cube, they follow the directions on the dot card with the five dots. They can roll the number cube twice to generate each of their fractions, once for the numerator and once for the denominator. Alternatively, they can pick a domino and use one side for the numerator and one for the denominator. The students then compare/order their fractions using one of the representations they rolled on the Representation Cube. They then check their ordering using the second representation they rolled.

Apply: If students roll "Apply," and they roll a six on a number cube, they follow the directions on the dot card with the six dots. They can roll the number cube twice to generate their fractions, once for the numerator and once for the denominator. Alternatively, they can pick a domino and use one side for the numerator and one for the denominator. The students then create a contextual word problem that models the action – in this case, apply.

Representation Cube Options

Fraction Model: Students create a fraction model. You might have them choose their model or assign a specific type such as an area model, a set model, a number line model, etc.

Symbolic/Numeric: Symbolic/numeric representation is the typical numerator over the denominator using a vinculum.

Concrete: For a concrete representation students might draw a picture, use manipulatives, cut paper, or use some other physical representation for the fractions.

Contextual: Students create a contextual situation that uses the fractions they rolled. This would be like the typical word problem.

Number Line: Students use a number line to represent the fractions' location as well as any actions they are performing on the fractions.

Verbal: Students can either simply write the fraction(s) in words, as they work through the actions, or they can be asked to verbally state the work they are doing using precise mathematical terminology.

Algebraic Reasoning

In this adaptation of a cubing and think dots activity, there are only two cubes; an operation cube and a representation cube. There are also a single-digit dodecahedron and a multiple of ten decahedrons that are provided for students to use. There are three sets of leveled think dot cards.

Students will either be working with the think dot cards and the dodecahedrons, or they will be working with the dodecahedrons and the cubes. The cards and cubes are not integrated as they are with the fraction activities.

Cubing Activity

Students roll the digits decahedron as many times as they need, based upon how many digits you want them to use for each of their numbers, as well as if they make only two numbers, or more. They then roll the operations cube to see what operation they use. They roll the representation cube to see what representation they use in their work. You can also include order of operations if you like by having them roll the operation cube more than once.

Operation Cube Options

Add: Students add the numbers they create.

Subtract: Students subtract the numbers they create.

Multiply: Students multiply the numbers they create.

Divide: Students divide the numbers they create.

Compare: Students compare or order the numbers they create based upon your directions.

Free choice: Students get to decide which operation they work with.

Representation Cube Options

Contextual situation: Students create a scenario in which the operation(s) apply.

Write the verbal form: Students use the number words and terms for operations as they describe what they did.

Concrete/Manipulatives: Students use concrete materials, manipulatives, etc. to represent their work.

Model partial products/quotients: Students use partial products/quotients in their work. If they are not using multiplication or division, they need to roll again.

Write the numeric form: Use numerals as they work.

Create an array: Students can either draw an array or use manipulatives to create an array as they work. If they are not using multiplication or division, they need to roll again.

Fractions

Action Cube

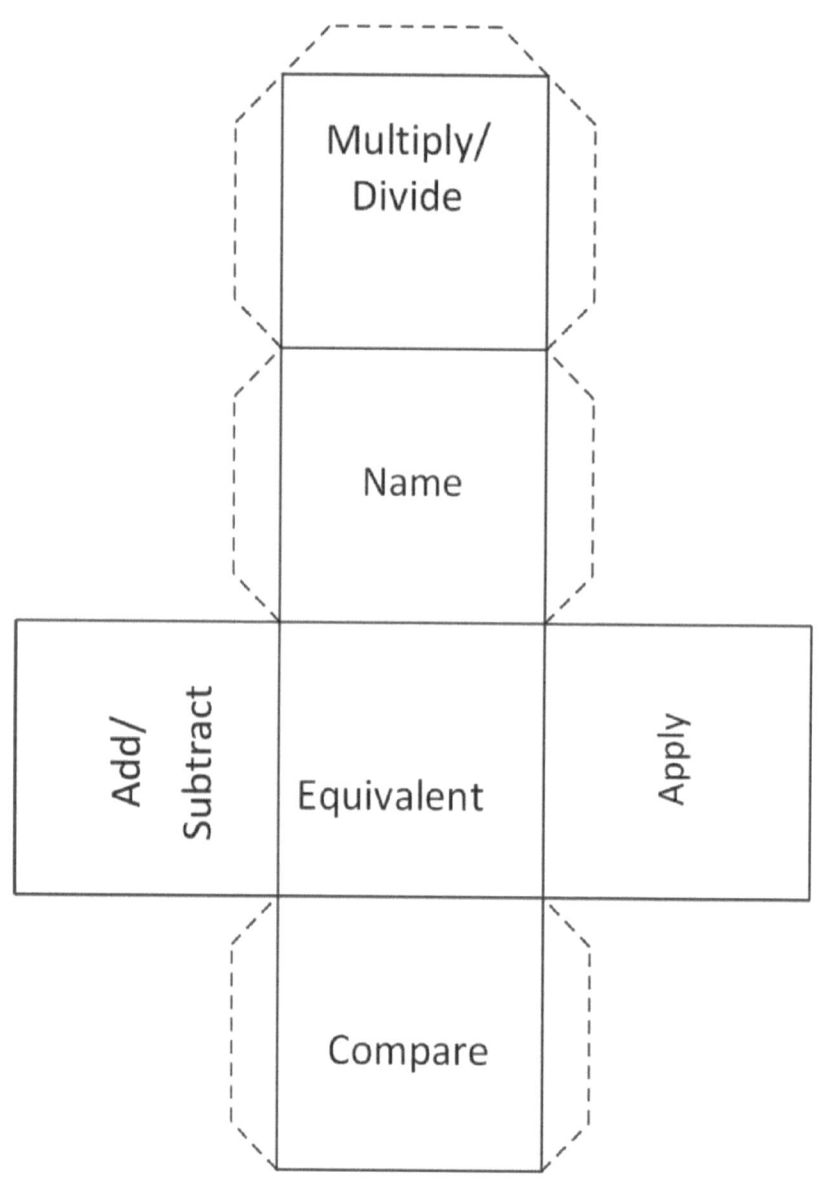

Fractions Representation Cube

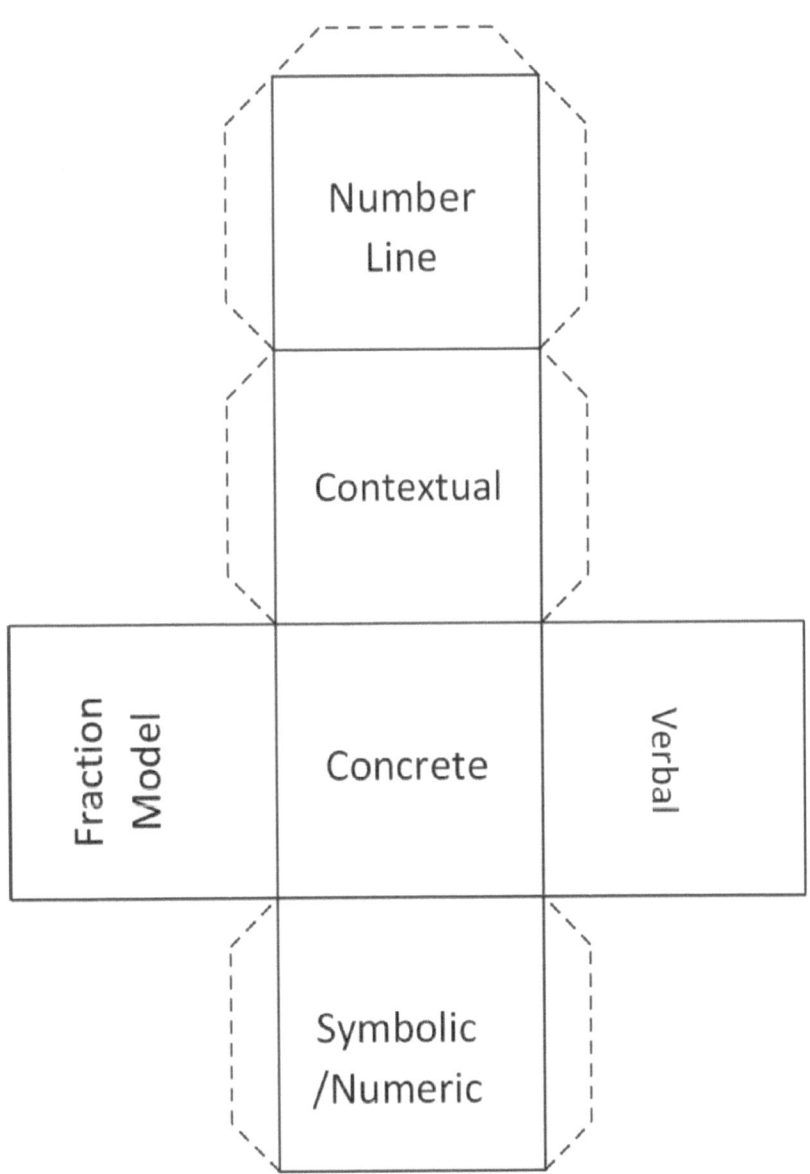

Think Dot Cards

Fractions

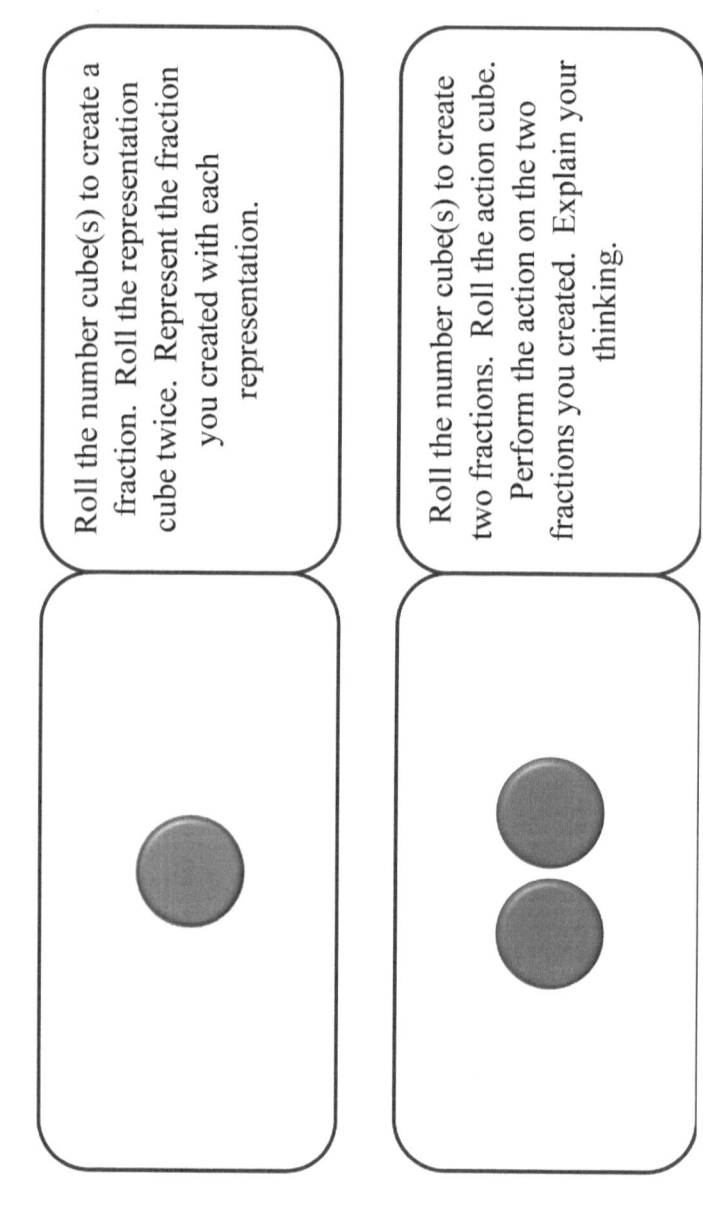

Roll the number cube(s) to create a fraction. Roll the representation cube twice. Represent the fraction you created with each representation.

Roll the number cube(s) to create two fractions. Roll the action cube. Perform the action on the two fractions you created. Explain your thinking.

Roll the number cube(s) to create two fractions. Roll the action cube. Choose a representation from the representation cube to use to perform the action. Explain your thinking.

Roll the action cube. Write a letter to a student who has been absent explaining how to do the action you rolled. You may use an example, if needed.

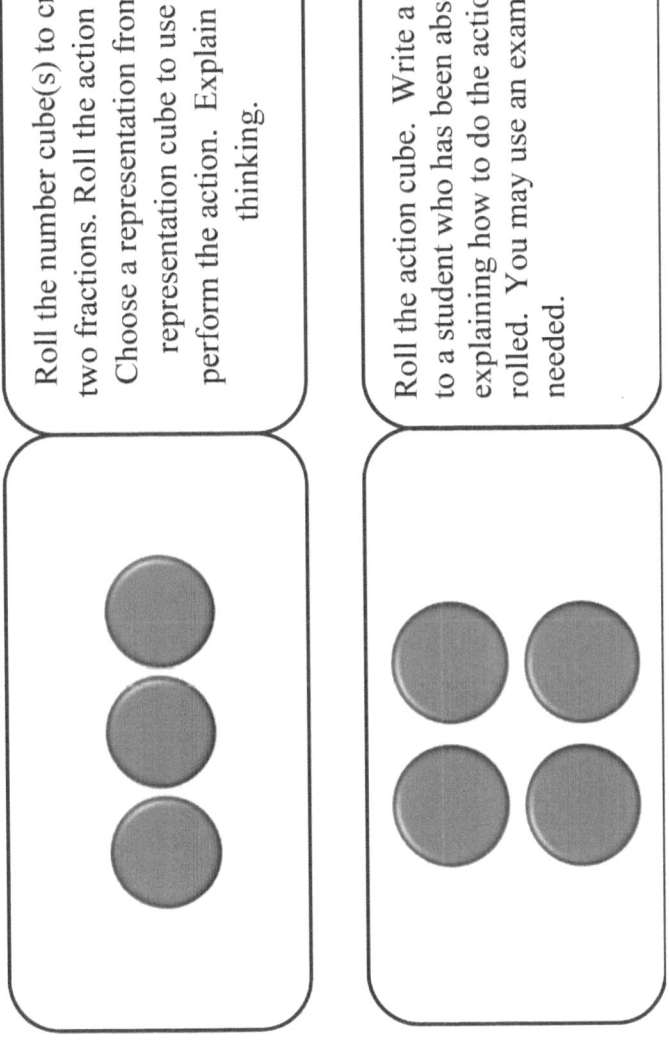

Roll the number cube(s) to create 5 fractions. Roll the representation cube twice. Order your fractions from greatest to least using one of the representations you rolled. Check using the second

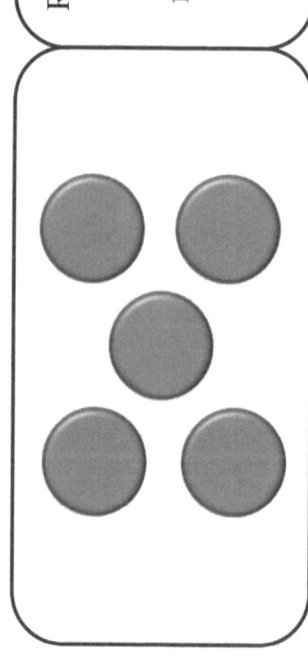

Roll the number cube(s) to create two fractions. Roll the action cube. Create a contextual word problem that models the action you rolled.

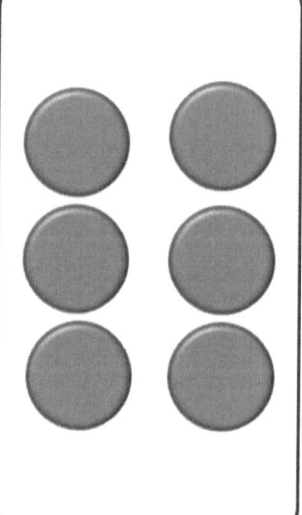

Algebraic Reasoning

Action Cube

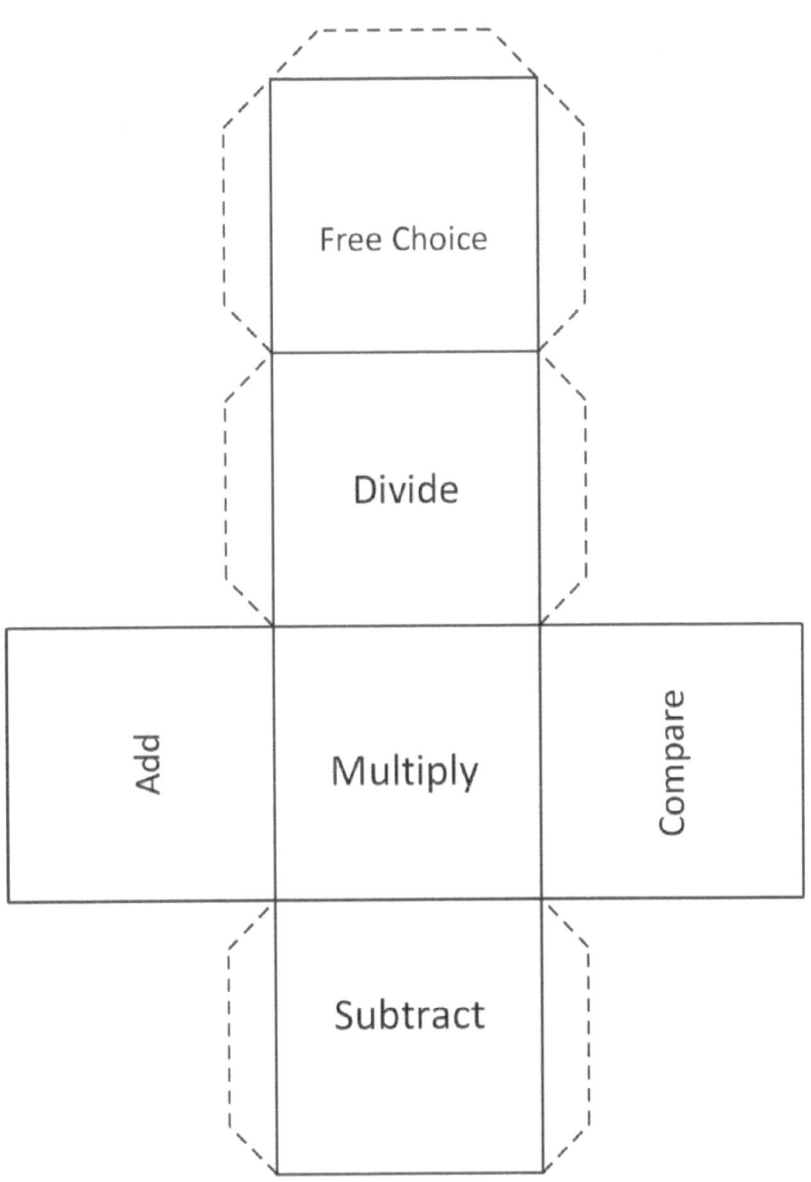

Algebraic Reasoning Representation Cube

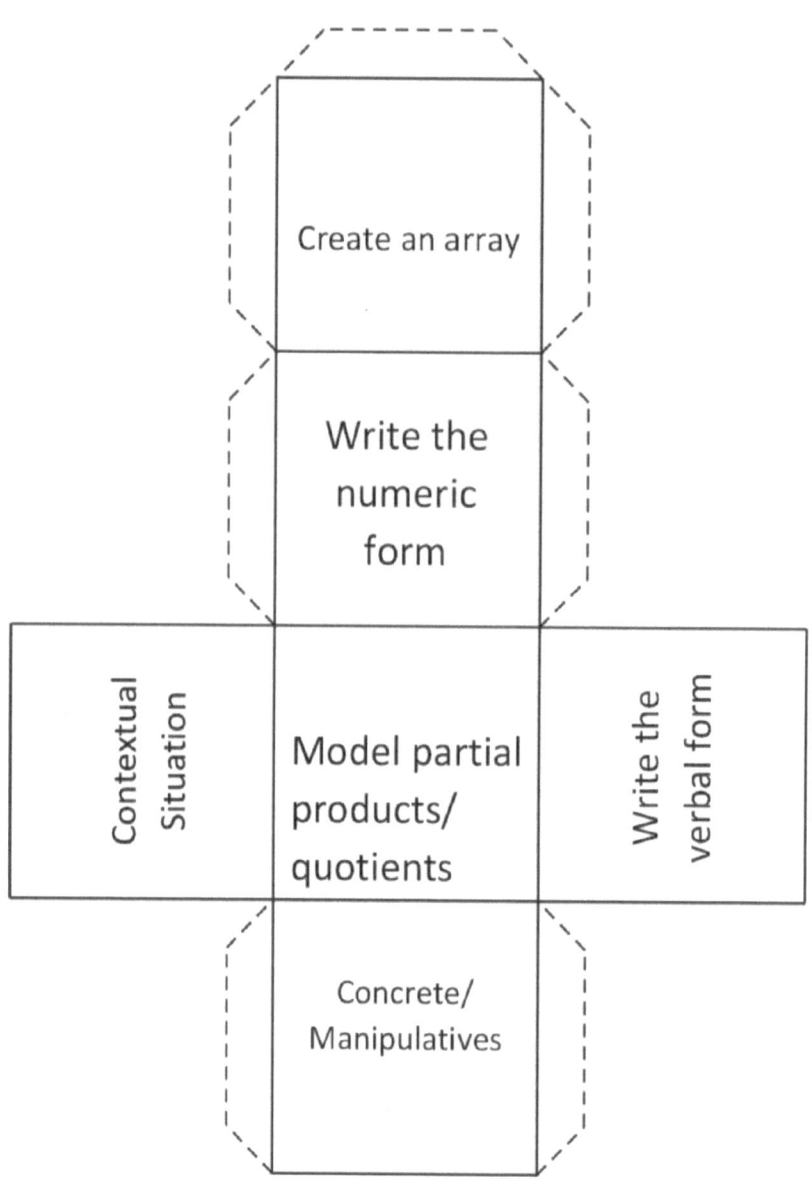

Algebraic Reasoning Digits Dodecahedron

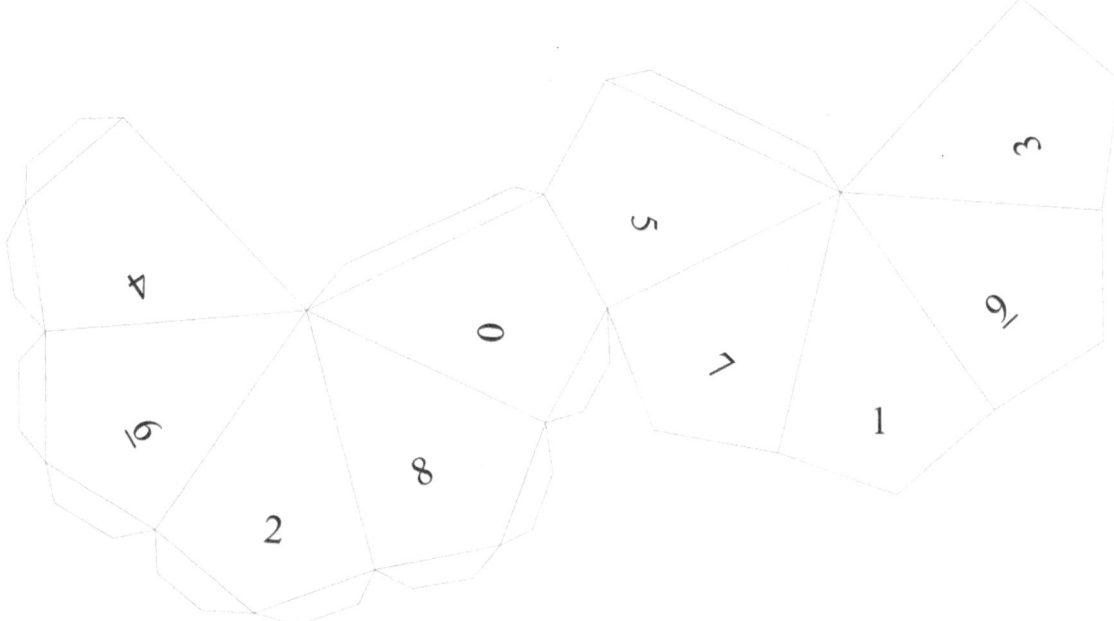

Algebraic Reasoning Multiples of Ten Dodecahedron

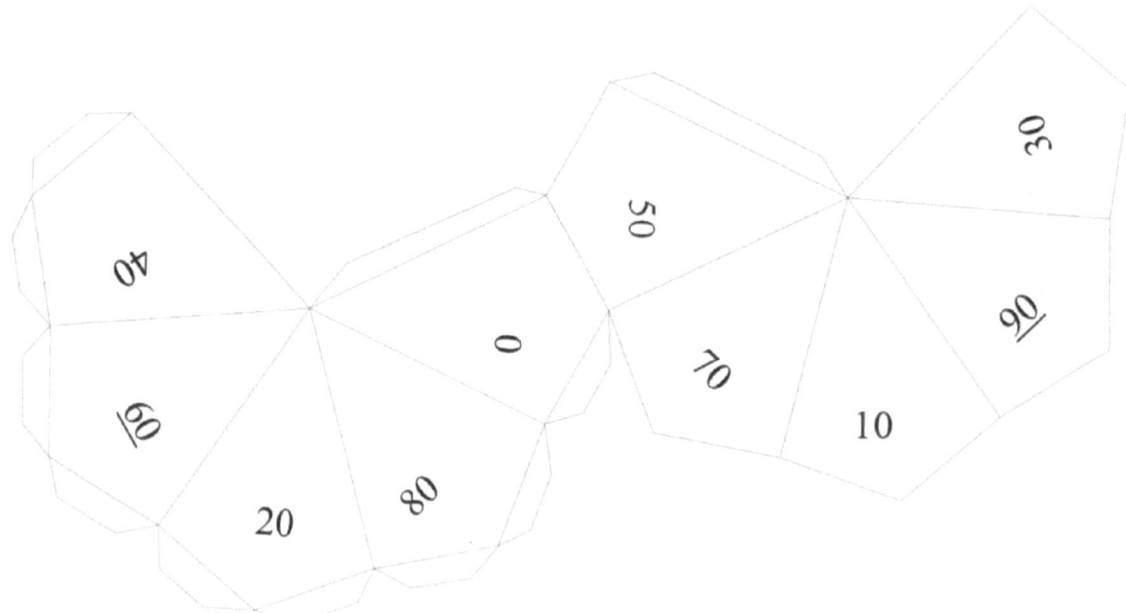

Think Dot Cards
Level 1 Algebraic Reasoning

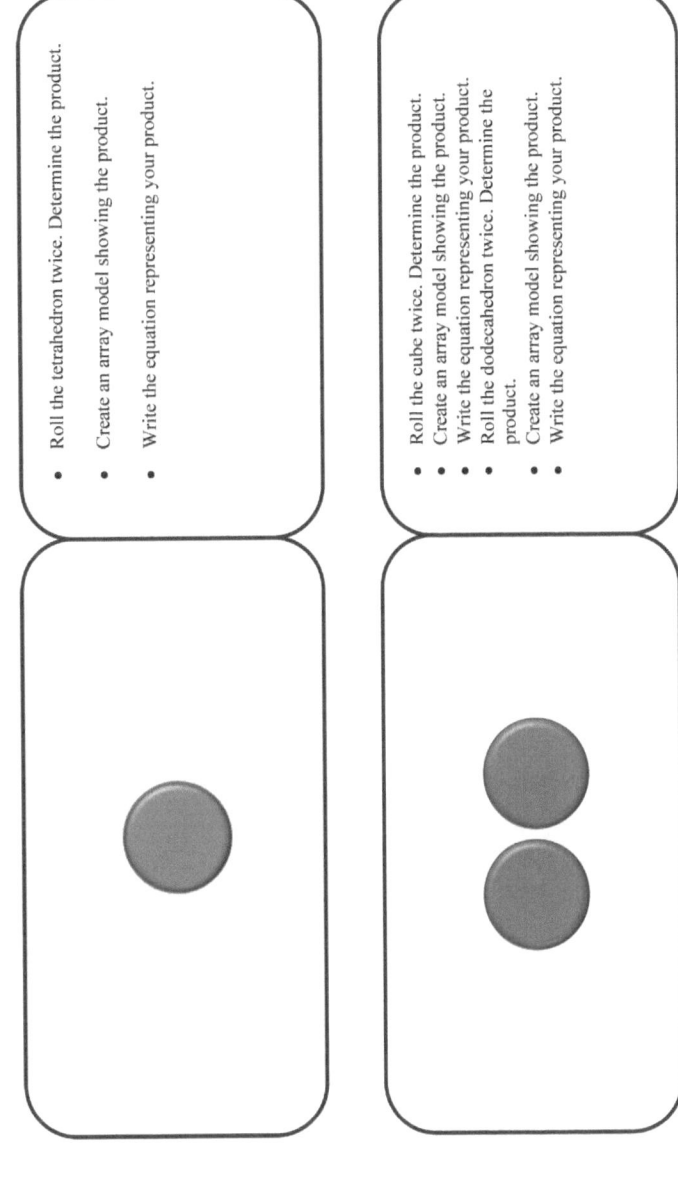

- Roll the tetrahedron twice. Determine the product.
- Create an array model showing the product.
- Write the equation representing your product.

- Roll the cube twice. Determine the product.
- Create an array model showing the product.
- Write the equation representing your product.
- Roll the dodecahedron twice. Determine the product.
- Create an array model showing the product.
- Write the equation representing your product.

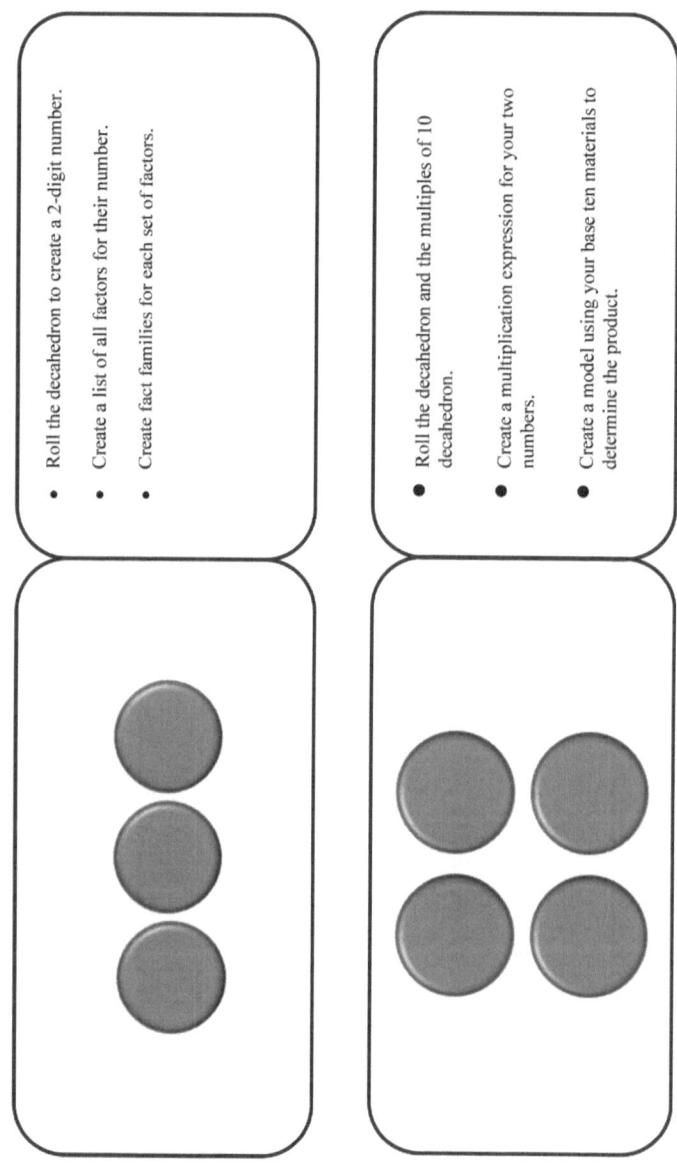

- Roll the decahedron to create a 2-digit number.
- Create a list of all factors for their number.
- Create fact families for each set of factors.

- Roll the decahedron and the multiples of 10 decahedron.
- Create a multiplication expression for your two numbers.
- Create a model using your base ten materials to determine the product.

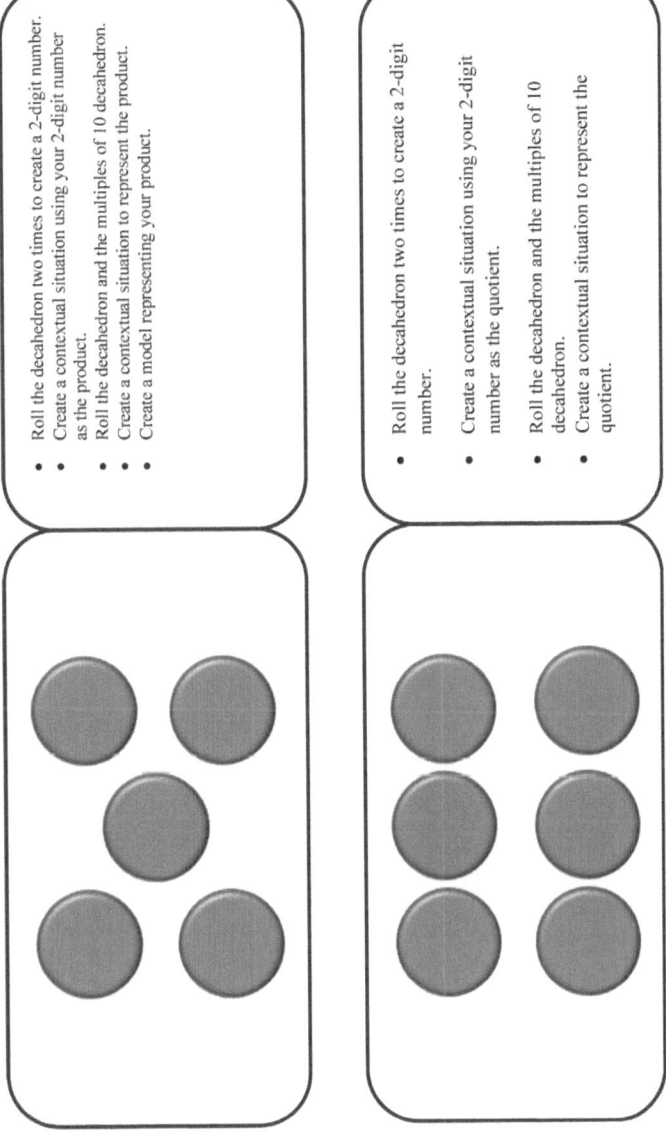

- Roll the decahedron two times to create a 2-digit number.
- Create a contextual situation using your 2-digit number as the product.
- Roll the decahedron and the multiples of 10 decahedron.
- Create a contextual situation to represent the product.
- Create a model representing your product.

- Roll the decahedron two times to create a 2-digit number.
- Create a contextual situation using your 2-digit number as the quotient.
- Roll the decahedron and the multiples of 10 decahedron.
- Create a contextual situation to represent the quotient.

Think Dot Cards

Level 2 Algebraic Reasoning

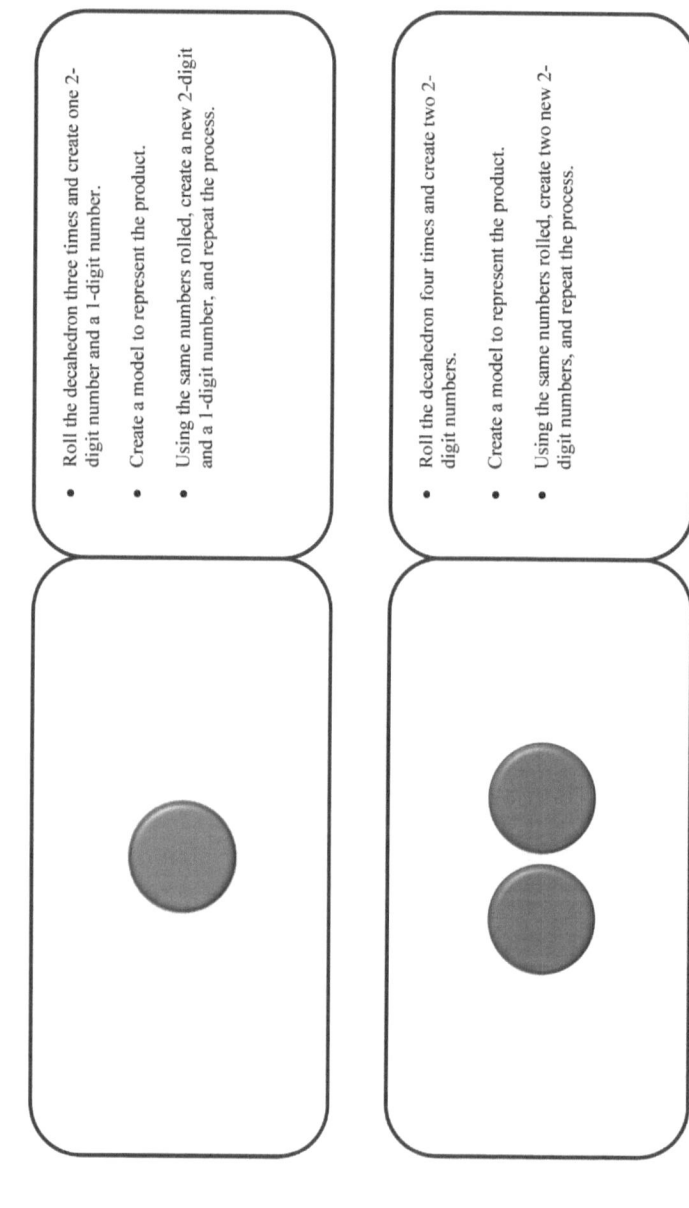

- Roll the decahedron three times and create one 2-digit number and a 1-digit number.

- Create a model to represent the product.

- Using the same numbers rolled, create a new 2-digit and a 1-digit number, and repeat the process.

- Roll the decahedron four times and create two 2-digit numbers.

- Create a model to represent the product.

- Using the same numbers rolled, create two new 2-digit numbers, and repeat the process.

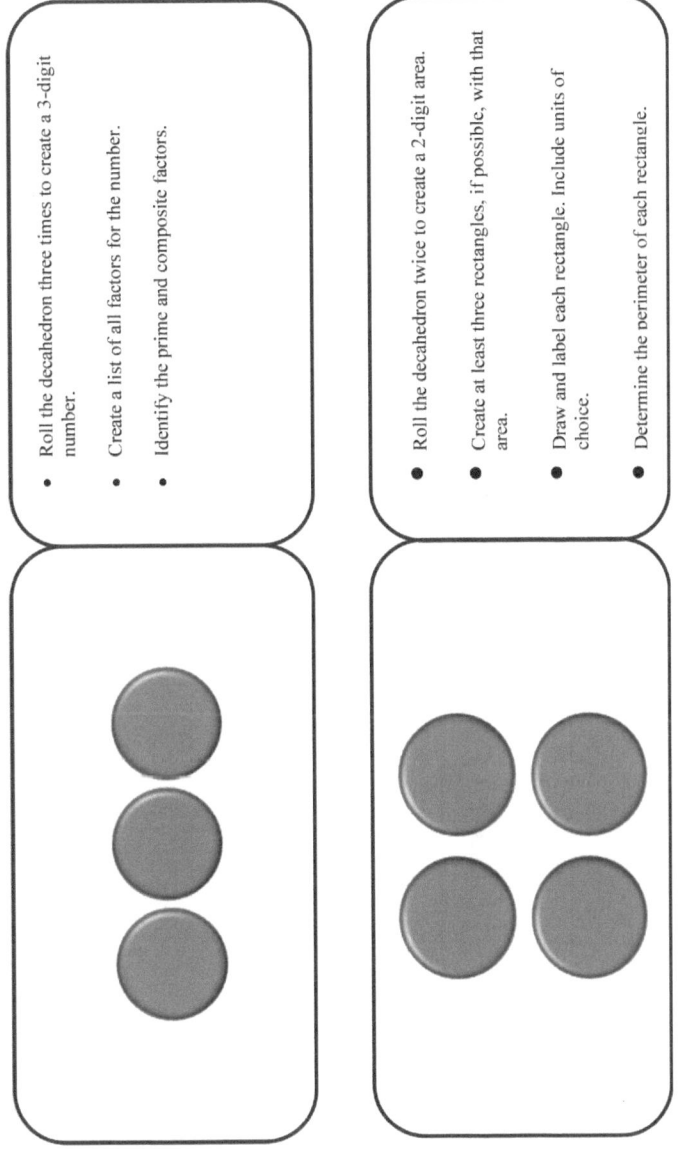

- Roll the decahedron three times to create a 3-digit number.

- Create a list of all factors for the number.

- Identify the prime and composite factors.

- Roll the decahedron twice to create a 2-digit area.

- Create at least three rectangles, if possible, with that area.

- Draw and label each rectangle. Include units of choice.

- Determine the perimeter of each rectangle.

- Roll the decahedron three times to create a 3-digit dividend.
- Roll the decahedron once to create a 1-digit divisor. Illustrate and explain the calculation.
- Roll the decahedron four times to create a 4-digit dividend.
- Roll the decahedron once to create a 1-digit divisor. Illustrate and explain the calculation.

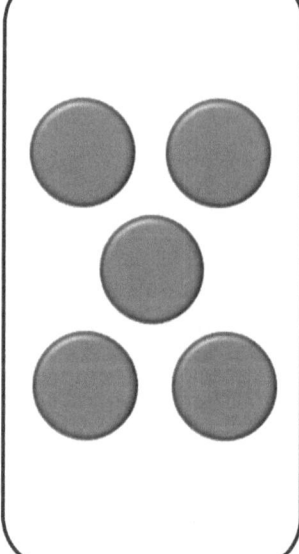

- Soccer teams are playing in a tournament. Each player has a _____ qt. (roll tetrahedron) water bottle. There are _____ (roll decahedron) players in the tournament. Each large container of sports drink holds _____ gal. (roll dodecahedron). How many sport drink containers are needed for each player to fill their bottles one time?

- Create another contextual problem like above. Roll, fill-in-the-blanks, and solve.

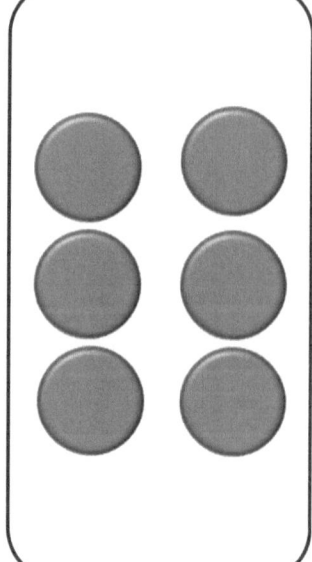

Think Dot Cards
Level 3 Algebraic Reasoning

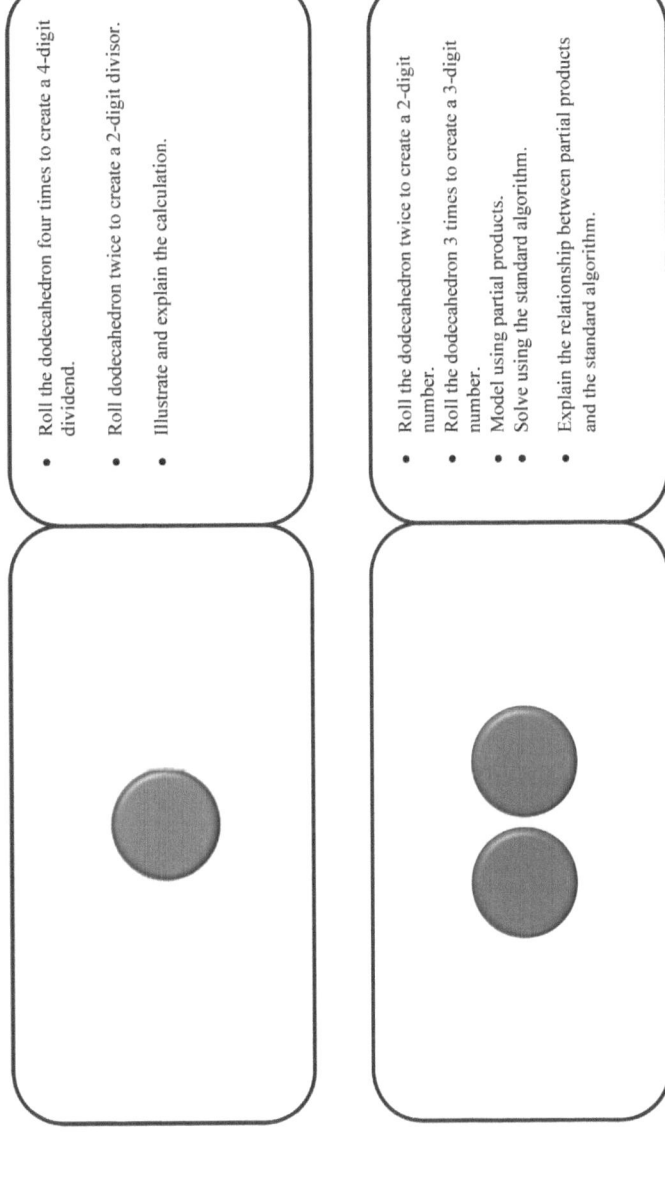

- Roll the dodecahedron four times to create a 4-digit dividend.
- Roll dodecahedron twice to create a 2-digit divisor.
- Illustrate and explain the calculation.

- Roll the dodecahedron twice to create a 2-digit number.
- Roll the dodecahedron 3 times to create a 3-digit number.
- Model using partial products.
- Solve using the standard algorithm.
- Explain the relationship between partial products and the standard algorithm.

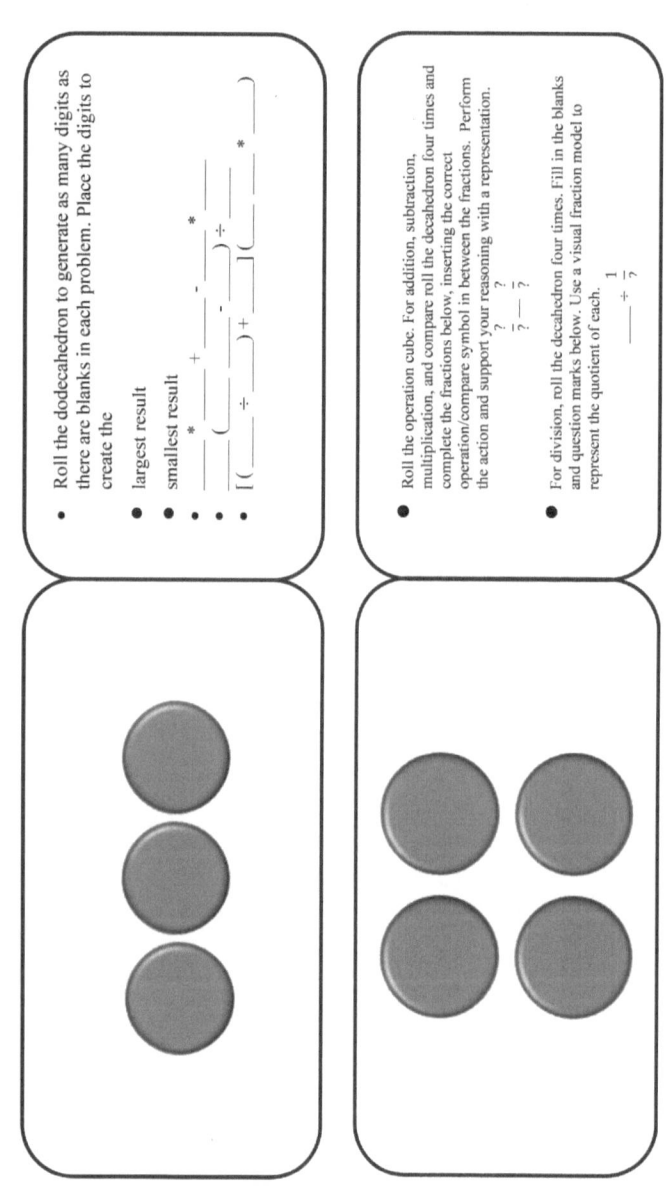

- Roll the dodecahedron to generate as many digits as there are blanks in each problem. Place the digits to create the
  - largest result
  - smallest result

$$[(\underline{\quad} \div \underline{\quad}) + \underline{\quad}] * \underline{\quad}$$

$$*\ \underline{\quad} - \underline{\quad} \div \underline{\quad}$$

$$(\underline{\quad} + \underline{\quad}) - \underline{\quad}\ *$$

- Roll the operation cube. For addition, subtraction, multiplication, and compare roll the decahedron four times and complete the fractions below, inserting the correct operation/compare symbol in between the fractions. Perform the action and support your reasoning with a representation.

$$\frac{?}{?} \quad \underline{\quad} \quad \frac{?}{?}$$

- For division, roll the decahedron four times. Fill in the blanks and question marks below. Use a visual fraction model to represent the quotient of each.

$$\underline{\quad} \frac{1}{?} \div \underline{\quad}$$

- Roll the decahedron three times and assign your results to _____ (length), _____ (width), and _____ (height)
- Roll the decahedron six times to create three fraction measures and assign your results to _____ (length), _____ (width), and _____ (height)
- Determine the volume of a rectangular prism given your dimensions. Draw and label each prism.

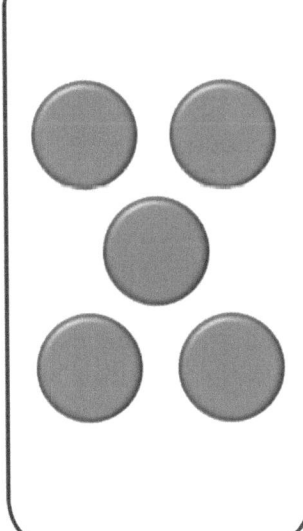

- Roll the multiples of ten decahedron and the decahedron. Calculate the sum and let this represent the volume of a rectangular prism.
- Roll the decahedron and this value represents one of the dimensions of the rectangular prism.
- Determine possible measurements of the other two dimensions and show your thinking. Draw and label the resulting rectangular prism.

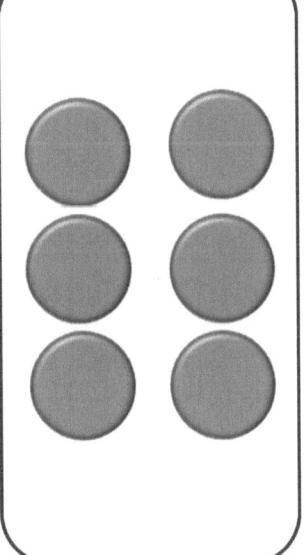

CHAPTER 10

RAFT

The RAFT writing strategy allows students to utilize their creativity within a mathematical context. Students are given a situation, along with a structure, which sets the parameters for the piece. If using as a formative assessment, the topic can be kept vague, as the bold text indicates, to determine whether students have internalized the key understanding(s) of the topic. If the intention is to ensure specific mathematical points are included in the piece, clearly state that expectation within the topic guidelines as the additional text demonstrates. This could be in the form of a rubric.

Role	Audience	Format	Topic
Fraction	Number Line	Directions	**Find My Location:** Explain how to locate a fraction on the number line.
Hundred Thousands	Tens	Sketch	**What Am I Worth?** Compare the value of a digit in tens place with the same digit in hundred-thousands' place.

DOI: 10.4324/9781003374572-12

Role	Audience	Format	Topic
Area of a rectangle	Arrays	Invitation to a Family Reunion	**How We Are Related:** Explain the relationship between the area of a rectangle and the arrays that model it.
Acute angle	Obtuse angle	Cartoon	**How Many** Create a viable argument for how many of each type of angle can be in a triangle.
Circle	Rectangle	Drawing	**Division of Assets:** Show the basic unit fractions partitioned in both a circle and a square.
Divisor	Division	Missing persons ad	**Looking For:** Discuss remainers in quotients.
One-quarter	One-eighth	Line plot	**Location, location, location!** Create a neighborhood of unit fractions on a line plot.
Expression	Equation	Partnership agreement	**Our Collaboration:** Discuss how expressions and equations work together.
Fraction	Operation Signs	Social media post	**Drawing Conclusions:** You don't always increase our value!
Sequence of numbers	Next Term	Instructions	**How do you Know?** Create a set of instructions for determining the next term in a sequence.
You	Absent classmate	Notes	**Irreconcilable Differences:** Explain the difference in the English standard system of measurement and the Metric system of measurement
Student Choice	Student Choice	Student Choice	Student Choice

TJ's Restaurant

Monday - Thursday: 10 am - 11 pm
Friday - Sunday: 9 am - midnight

Shareables

Nachos	$9.95

House-made crispy tortilla chips, pepper-jack, jalapenos, black olives, fresh cut salsa, sour cream, guacamole

Mozzarella sticks (6)	$8.50

Served with marinara sauce

Chicken wings (8)	$12.95

Plain or sauce of your choice: BBQ, Honey Mustard, Ranch

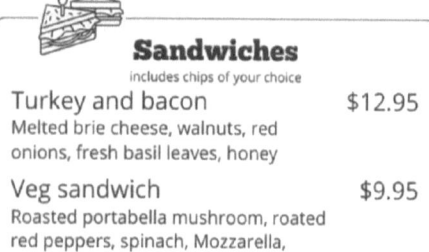

Sandwiches
includes chips of your choice

Turkey and bacon	$12.95

Melted brie cheese, walnuts, red onions, fresh basil leaves, honey

Veg sandwich	$9.95

Roasted portabella mushroom, roated red peppers, spinach, Mozzarella, tomato slices, red onions

Pizza

3 - Cheese 8"	$9.95

Mozzarella, feta, asiago with balsamic drizzle

Meats for all 12"	$19.95

Pepperoni, Italian sausage, ham, bacon, mozzarella, red sauce

Veggie Delite 8"	$9.95

Tomato, mushrooms, sweet pepper, red onions, spinach, mozzarella, red sauce

White pizza 8"	$12.95

Chicken, spinach, red onion, white sauce, asiago cheese

Fire pizza 12"	$19.95

Italian sausage, pepperoni, pepperoncini, mushrooms, red onions, habanero pepper, mozzarella, BBQ sauce

Hot Dogs
includes chips of your choice

All American	$4.50
Polish Dog	$5.50

Sauerkraut, spicy mustard

Spicy Hot Dog	$7.50

With corn chips and fresh chilies

Burgers
includes fries

House	$10.75

Angus beef burger topped with tomato, red onion, lettuce, American cheese

All American Cheeseburger	$13.00

2 beef patties, cheddar cheese, grilled onions, pickles, lettuce, ketchup, mustard

Black Bean	$11.75

Black bean patty, lettuce, onion, tomato, pickle, sour cream, guacamole

Turkey Burger	$12.50

Ground turkey patty, tomato, sweet onion, remoulade sauce, swiss cheese

Wraps

Chicken	$6.25

Chicken, cheddar cheese, tomato, lettuce, ranch

Roast Beef	$9.00

Shaved roast beef, provolone, crispy onions, BBQ sauce

Veg	$7.25

Roasted portabello mushroom, tomato, spinach, garlic sauce, parmesan cheese

Here is an example of one way that you can "RAFT-ize" a common topic. The set-up below shows how you can take a menu from a local eatery and use it to let the students work with basic computation and money. Although not computed here, discussions need to be facilitated around taxes, tips, etc. so they understand this may not be the total cost of the meal.

Scenario: In your job as a reporter, you have been tasked with reporting on a new local restaurant for the "Dining on a Budget" segment of your local newscast. Given the menu, you need to offer several options for a family of four to eat a meal under $50 and under $100.

Role	What is the writer's role?	Dining on a Budget News Reporter
Audience	Who will be reading the piece?	The TV Audience
Format	What is the best way to present the information?	News report – including graphics
Topic	Who or what is the subject?	Menu options for the budget conscious

Additional Format Ideas

Speech	Journal Entry	Script	Song
Public service announcement	Brochure	Story board	Advice column
Text message	Petition	Letter (apology, persuasive, thank you, complaint)	Commercial
Advertisement	Sticker	Campaign speech	Wanted poster
Itinerary	Personal Ad	Nursery Rhyme/Riddle	Infographic
Editorial	Eulogy	Current event	Debate

Additional Samples

Role: Numerator
Audience: Denominator
Format: Recipe
Topic: Part to Whole (include discussion of doubling a recipe)

Role: Quadrilateral
Audience: Trapezoid
Format: Family Tree

Topic: Where do I belong?

Role: Mixed Numbers
Audience: Rational Numbers
Format: Trading Card
Topic: Multiple Representations

Role: Clock
Audience: Students in School
Format: Schedule
Topic: How Much Time? (for example: time between classes, how much time until..., how much time since...)

CHAPTER 11

Question Quilt

The question quilt can be a strategy for differentiation that allows for student agency as they read and think about questions and statements relating to a topic of study. They then choose a few of the boxes to think about further. Students decide if they agree or disagree with the statement(s) and/or answer the question(s) and write justification supporting their responses. Questions and statements can be framed to accommodate a variety of levels of learners. Some of the questions/statements are more geared to the lower grades in the grade band as well as some are more geared for the upper grades.

Another option is to give students the question quilt as you are beginning a topic, and they can discuss the questions as they progress through the development of the topic.

Sample directions: Choose at least three questions or statements from the question quilt. Answer the question or decide if you agree or disagree with the statement. Justify your responses fully.

DOI: 10.4324/9781003374572-13

Question Quilt
Fractions

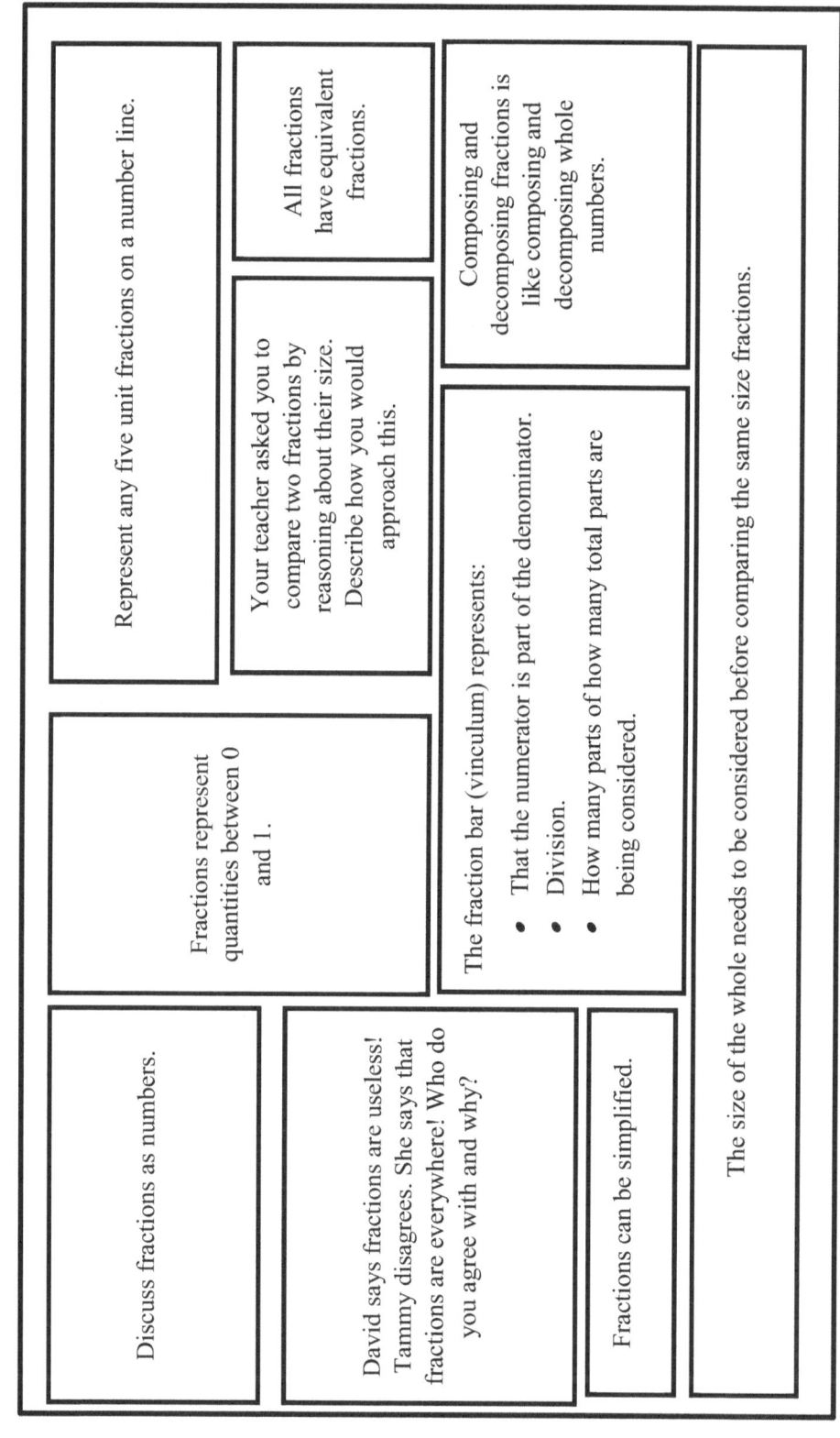

Represent any five unit fractions on a number line.

All fractions have equivalent fractions.

Your teacher asked you to compare two fractions by reasoning about their size. Describe how you would approach this.

Composing and decomposing fractions is like composing and decomposing whole numbers.

Fractions represent quantities between 0 and 1.

The fraction bar (vinculum) represents:

- That the numerator is part of the denominator.
- Division.
- How many parts of how many total parts are being considered.

Discuss fractions as numbers.

David says fractions are useless! Tammy disagrees. She says that fractions are everywhere! Who do you agree with and why?

Fractions can be simplified.

The size of the whole needs to be considered before comparing the same size fractions.

Question Quilt
Algebraic Reasoning

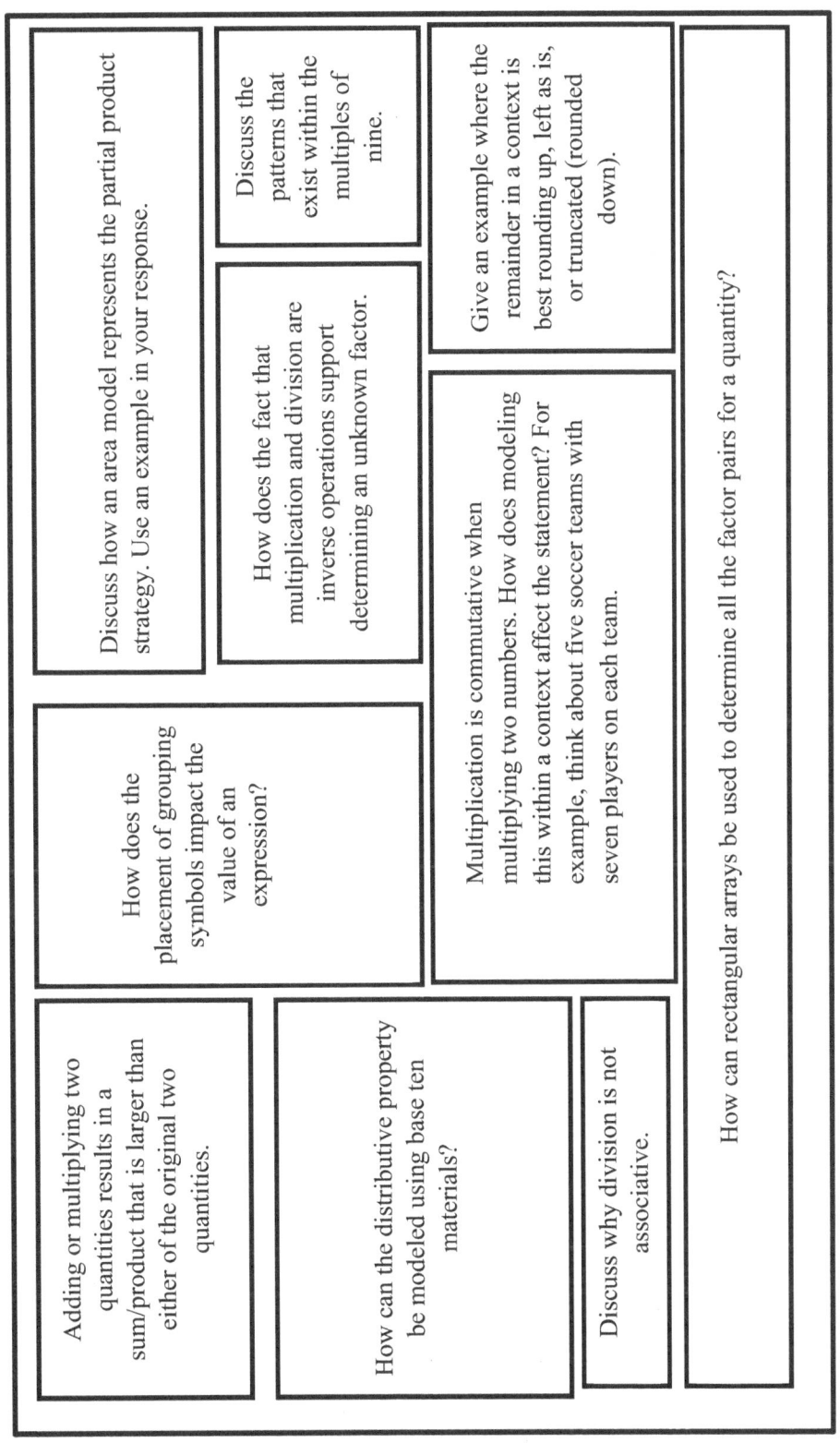

Question Quilt
Measurement and Units

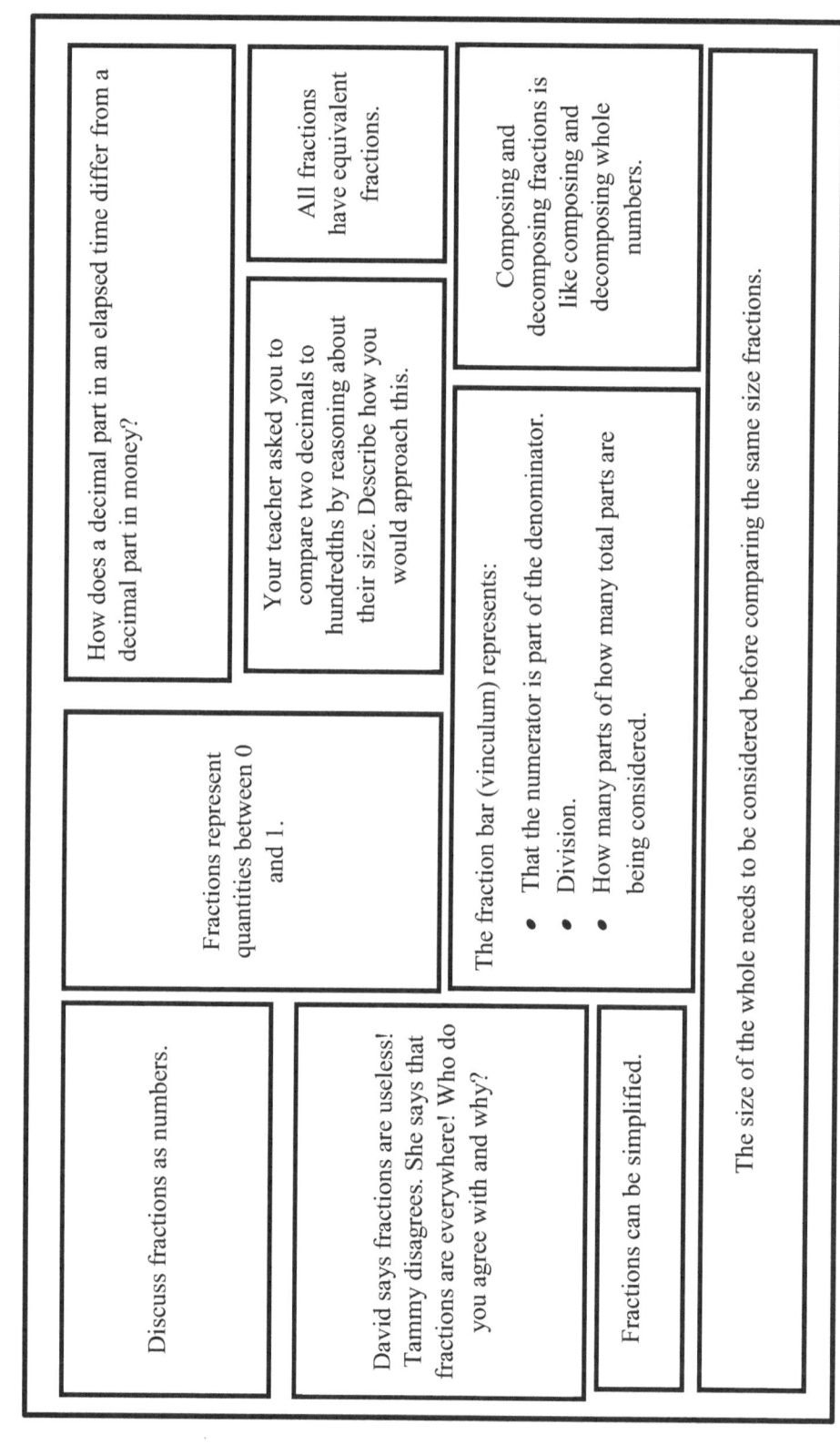

How does a decimal part in an elapsed time differ from a decimal part in money?

All fractions have equivalent fractions.

Your teacher asked you to compare two decimals to hundredths by reasoning about their size. Describe how you would approach this.

Composing and decomposing fractions is like composing and decomposing whole numbers.

Fractions represent quantities between 0 and 1.

The fraction bar (vinculum) represents:

- That the numerator is part of the denominator.
- Division.
- How many parts of how many total parts are being considered.

Discuss fractions as numbers.

David says fractions are useless! Tammy disagrees. She says that fractions are everywhere! Who do you agree with and why?

Fractions can be simplified.

The size of the whole needs to be considered before comparing the same size fractions.

CHAPTER 12

Always, Sometimes, and Never

Using always, sometimes, and never questions provides the opportunity for students to investigate a statement and determine whether it is true all the time, some of the time, or never. Typically, always, sometimes, and never have been associated with geometry, but as you can see from the following pages, these are as easily adapted to other topics. They are rich tasks because they engage students in a higher level of reasoning and communication than typical questions might. This activity allows teachers to look into the thinking of the students.

For upper elementary students, it is suggested that you work through a couple of examples with the class so they understand this question type and what is expected of them as write their answers. Students may not simply write A, S, or N. They must validate and support their choice. If they determine the statement is sometimes, students should be required to offer a case for when the statement does hold (an example) as well as when it does not (a counter example). Many true and false questions can easily be adapted for always, sometimes, and never statements.

In the following, the topical examples start out at a lower level and build to the higher level. You can pick and choose based upon your grade level and then needs of your students.

DOI: 10.4324/9781003374572-14

Sample Statements

Topic 1: Number and Quantity

Place Value

1. Multiplying a number by 10 does not change the one's place of the first number.
2. Multiplying a number by 10 does not change the one's digit of the first number.
3. Expanded form of a number models the value of each of the digits in the number.
4. When you add ten to a number, the sum is a multiple of ten.
5. A decimal written to the thousandths place is larger than a number written to the hundredths place.

Fractions

1. Three halves can be represented on a number line.
2. Non-unit fractions can be decomposed into equal partitions.
3. Equivalent fractions have common denominators.
4. The numerator describes how many parts of a whole are being considered.
5. The numerator of a fraction is less than the denominator.
6. One-fourth is greater than one-third.
7. Fractions can be simplified.
8. Fractions represent quantities between 0 and 1.
9. If the numerator is larger than the non-zero denominator, the number is not a fraction.
10. All whole numbers can be expressed as a fraction.

Exponents

1. Numbers can be represented using a base number and an exponent for powers of ten.
2. In a power of ten, the exponent above the ten indicates how many places the decimal point is moving.
3. The number of zeros in a product is equal to the whole number exponent in a power of ten.
4. The greater the exponent, the greater the quantity represented.
5. Any quantity raised to the "0" power is "0."

6. Quantities raised to even powers of ten are even.
7. Exponents represent powers of 10.
8. Multiplying a number by 10 does not change the one's place of the original number.

Topic 2: Algebraic Reasoning

Operating with Numbers

Fractions

1. Any number over itself in a fraction is equivalent to one.
2. "0" over any number, in a fraction, is "0."
3. The size of the whole needs to be considered before comparing the same-size fractions.
4. All fractions have equivalent fractions.
5. Fractions represent quantities between 0 and 1.
6. The product, when multiplying two fractions, is larger than the two factors.
7. When adding or subtracting two fractions, you add/subtract the numerators and you add/subtract the denominators.
8. You can take half of a number.
9. When multiplying a fraction by its reciprocal, the resulting product is the identity element for multiplication.
10. $\dfrac{a}{b} = \dfrac{n \cdot a}{n \cdot b}$, if $b \neq 0$.

Models/Equations

1. Parentheses are needed when adding two or more numbers.
2. Parentheses are needed when dividing two or more numbers.
3. Equations are statements showing equality between two expressions.
4. Equations have an unknown.
5. Equations have equal signs.
6. All equations are expressions.
7. A rectangular array is a visual model of an equation.
8. All equations can be represented using area models.
9. The equal sign is used to compare fractions.
10. Contextual situations are represented by equations.

Topic 3: Geometric Reasoning

Family of Quadrilaterals

1. If two rectangles have the same perimeter, they have the same area.
2. Rectangles are squares.
3. Squares are rectangles.
4. If two squares have the same area, they have the same perimeter.
5. A rhombus is a trapezoid.
6. Quadrilaterals have right angles.
7. Parallelograms have no sides congruent.
8. Quadrilaterals do not have obtuse angles.
9. Quadrilaterals have at least one set of parallel sides.
10. Quadrilaterals can be decomposed into triangles.

Triangles

1. A scalene triangle can be isosceles.
2. An equilateral triangle can be equiangular.
3. Triangles can have more than one obtuse angle.
4. Right triangles are isosceles.
5. Acute triangles are right triangles.
6. Scalene triangles have two acute angles.
7. A triangle can have two right angles.
8. Triangles have no parallel sides.
9. Any angle 90° or less is a right angle.
10. When a rectangle is partitioned along one of its diagonals, the resulting triangles are congruent.

Angles, Lines, and Symmetry

1. Parallel lines are perpendicular.
2. Three lines can be parallel.
3. The length of a line can be measured in inches.
4. All figures have lines of symmetry.
5. A triangle has exactly two lines of symmetry.
6. A quadrilateral has four lines of symmetry.
7. An obtuse, right triangle can be drawn.
8. A scalene, right triangle can be drawn.
9. A figure can have a perimeter and an area that are equal.
10. An obtuse angle has a measure greater than a right angle.

Topic 4: Measurement and Units

Geometric Measurements

1. One-inch measurement is greater than one-centimeter measurement.
2. To convert a measurement to millimeters, multiply by a thousand.
3. Units of measurements for perimeter are the same as units for measurement for area.
4. Angles are measured in inches.
5. Two rectangles have the same area but different perimeters.
6. Protractors can be used to measure angles.
7. Metric measure is base ten.
8. Decomposition of shapes aids in computing the perimeter and area for a complex figure.
9. Perimeter, area, and volume are related measurements.
10. The units used when labeling the volume for a figure are square units.

Non-geometric Measurement (Elapsed Time and Money)

1. Time measurements can be modeled on a number line.
2. Elapsed time is measured in minutes.
3. When thinking about time, 3.5 hours is the same as 3 hours and 5 minutes.
4. The decimal part in an elapsed time measure is different from the decimal part when representing money.
5. If I have "at least" $7.00, I might have $6.50.
6. When converting years to days, always multiply by 365.
7. The decimal fraction, $\frac{147}{100}$, represents 147 pennies, or $1.47.
8. When converting from cents to dollars, you multiply by $\frac{1}{100}$.
9. If it's 3 pm in New York, it is 3 pm in Tennessee.
10. Leslie has $27.59. If Leslie rounds the amount of money she has, it is $28.

Topic 5: Data Analysis, Probability, and Statistics

1. Dot plots and pictographs show the same information about a data set.
2. Reading a pictograph helps answer the question, "How many?"
3. Labels on a data set represent units of geometric measurement.
4. When creating a bar graph, a title is sufficient.
5. Addition and subtraction of fractions can be solved using the information displayed in a line plot.

6. If the data set represents several million, the data display should be divided into at least one-million sections.
7. Line plots only show one line per graph.
8. Data displays show comparisons.
9. Graphs are visual representations of data.
10. When creating a visual representation for a data set, the type of graph does not matter.

Sample Statements with Answers

Topic 1: Number and Quantity

Place Value

1. Multiplying a number by 10 does not change the one's place of the first number. **(A)**
2. Multiplying a number by 10 does not change the one's digit of the first number. **(S, if the one's digit was originally 0.)**
3. Expanded form of a number models the value of each of the digits in the number. **(A)**
4. When you add ten to a number, the sum is a multiple of ten. **(S)**
5. A decimal written to the thousandths place is larger than a number written to the hundredths place. **(S, for example, 0.200 and 0.20.)**

Fractions

1. Three-halves can be represented on a number line. (A) [Note: This reinforces students' understanding that a number line represents more than whole numbers and can extend beyond 1.]
2. Non-unit fractions can be decomposed into equal partitions. (A) [Note: For example, students should realize that $\frac{3}{4}$ can be partitioned into say, 6 equal parts just as 1 whole can.]
3. Equivalent fractions have common denominators. **(N)** [**Note: The case of a fraction being equivalent to itself is not considered here.**]
4. The numerator describes how many parts of a whole are being considered. (A) [Note: $\frac{5}{4}$ of a candy bar is one whole candy bar plus an additional fourth of the same type of candy bar.]
5. The numerator of a fraction is less than the denominator. **(S)**
6. One-fourth is greater than one-third. **(S)** [**Note: Depending on the size of the original whole.**]

7. Fractions can be simplified. **(S)**
8. Fractions represent quantities between 0 and 1. **(S)**
9. If the numerator is larger than the non-zero denominator, the number is not a fraction. **(N)**
10. All whole numbers can be expressed as a fraction. **(A)**

Exponents

1. Numbers can be represented using a base number and an exponent for powers of ten. **(A)**
2. In a power of ten, the exponent above the 10 indicates how many places the decimal point is moving. **(A)**
3. The number of zeros in a product is equal to the whole number exponent in a power of ten. **(S)** [**Note: For example,** $2 \times 10^2 = 200$, **this is true . However,** $3.04 \times 10^2 = 304$, **this is not true, as there are no zeros at the end of the product.**]
4. The greater the exponent, the greater the quantity represented. **(S)**
5. Any quantity raised to the "0" power is "0." **(N)**
6. Quantities raised to even powers of ten are even. **(S)**
7. Exponents represent powers of 10. **(S)** [**Note: This statement can be used to foreshadow what students will do in middle school when working with integer exponents.**]
8. Multiplying a number by 10 does not change the one's place of the original number. **(S)** [**Note: If your original factor was a single-digit number.**]

Topic 2: Algebraic Reasoning

Operating with Numbers

Fractions

1. Any number over itself in a fraction is equivalent to one. **(S, because of 0.)**
2. "0" over any number, in a fraction, is "0." **(S, can't have 0 in the denominator at this level.)**
3. The size of the whole needs to be considered before comparing the same size fractions. **(A)**
4. All fractions have equivalent fractions. **(A)**
5. Fractions represent quantities between 0 and 1. **(S, you can have quantities greater than 1 as well as less than 0.)**
6. The product, when multiplying two fractions, is larger than the two factors. **(S)**

7. When adding or subtracting two fractions, you add/subtract the numerators and you add/subtract the denominators. **(N)**
8. You can take half of a number. **(A)**
9. When multiplying a fraction by its reciprocal, the resulting product is the identity element for multiplication. **(A)**
10. $\dfrac{a}{b} = \dfrac{n \cdot a}{n \cdot b}$, if $b \neq 0$. **(A)**

Models/Equations

1. Parentheses are needed when adding two or more numbers. (N) [Note: Addition and multiplication are both commutative.]
2. Parentheses are needed when dividing two or more numbers. **(A)** **[Note: Subtraction and division are both not commutative.]**
3. Equations are statements showing equality between two expressions. **(A)**
4. Equations have an unknown. **(S)**
5. Equations have equal signs. **(A)**
6. All equations are expressions. **(N)**
7. A rectangular array is a visual model of an equation. **(A)**
8. All equations can be represented using area models. **(S)**
9. The equal sign is used to compare fractions. **(S)**
10. Contextual situations are represented by equations. **(S)**

Topic 3: Geometric Reasoning

Family of Quadrilaterals

1. If two rectangles have the same perimeter, they have the same area. **(S)**
2. Rectangles are squares. **(S)**
3. Squares are rectangles. **(A)**
4. If two squares have the same area, they have the same perimeter. **(A)**
5. A rhombus is a trapezoid. **(N)**
6. Quadrilaterals have right angles. **(S)**
7. Parallelograms have no sides congruent. **(N)**
8. Quadrilaterals do not have obtuse angles. **(S)**
9. Quadrilaterals have at least one set of parallel sides. **(S)**
10. Quadrilaterals can be decomposed into triangles. **(A)**

Triangles

1. A scalene triangle can be isosceles. **(S)**
2. An equilateral triangle can be equiangular. **(A)**

3. Triangles can have more than one obtuse angle. (**N**)
4. Right triangles are isosceles. (**S**)
5. Acute triangles are right triangles. (**N**)
6. Scalene triangles have two acute angles. (**S**)
7. A triangle can have two right angles. (**N**)
8. Triangles have no parallel sides. (**A**)
9. Any angle 90° or less is a right angle. (**S**)
10. When a rectangle is partitioned along one of its diagonals, the resulting triangles are congruent. (**A**)

Angles, Lines, and Symmetry

1. Parallel lines are perpendicular. (**N**)
2. Three lines can be parallel. (**S**)
3. The length of a line can be measured in inches. (**N**) [**Note: A line segment can be measured, but not a line.**]
4. All figures have lines of symmetry. (**S**)
5. A triangle has exactly two lines of symmetry. (**N**)
6. A quadrilateral has four lines of symmetry. (**S**)
7. An obtuse, right triangle can be drawn. (**N**)
8. A scalene, right triangle can be drawn. (**S**)
9. A figure can have a perimeter and an area that are equal. (**S**)
10. An obtuse angle has a measure greater than a right angle. (**A**)

Topic 4: Measurement and Units

Geometric Measurements

1. One-inch measurement is greater than one-centimeter measurement. (**A**)
2. To convert a measurement to millimeters, multiply by a thousand. (**S**)
3. Units of measurements for perimeter are the same as units for measurement for area. (**N**) [**Note: Perimeter is a one-dimensional measurement when computed and area is a two-dimensional measurement when computed. So, perimeter might be 4 cm, but area would be 4 cm².**]
4. Angles are measured in inches. (**N**)
5. Two rectangles have the same area but different perimeters. (**S**)
6. Protractors can be used to measure angles. (**A**)
7. Metric measure is base ten. (**A**)
8. Decomposition of shapes aids in computing the perimeter and area for a complex figure. (**S**)

9. Perimeter, area, and volume are related measurements. **(A)**
10. The units used when labeling the volume for a figure are square units. **(N)**

Non-geometric Measurement (Elapsed Time and Money)

1. Time measurements can be modeled on a number line. **(A)**
2. Elapsed time is measured in minutes. **(S)**
3. When thinking about time, 3.5 hours is the same as 3 hours and 5 minutes. **(N)**
4. The decimal part in an elapsed time measure is different from the decimal part when representing money. **(A)** [**Note: Students often do not realize that time is base 60 and money is base 10.**]
5. If I have "at least" $7.00, I might have $6.50. **(N)**
6. When converting years to days, always multiply by 365. **(S)** [**Note: Students need to think about how to deal with a leap year.**]
7. The decimal fraction, $\dfrac{147}{100}$, represents 147 pennies, or $1.47. **(A)**
8. When converting from cents to dollars, you multiply by $\dfrac{1}{100}$. **(A)**
9. If it's 3 pm in New York, it is 3 pm in Tennessee. **(S)** [**Note: Tennessee is one of several states that has two time zones. It has both Eastern and Central time zones.**]
10. Leslie has $27.59. If Leslie rounds the amount of money she has, it is $28. **(S)** [**Note: If rounding to dollars, yes. If rounding to cents, no.**]

Topic 5: Data Analysis, Probability, and Statistics

1. Dot plots and pictographs show the same information about a data set. **(A)**
2. Reading a pictograph helps answer the question, "How many?" **(A)**
3. Labels on a data set represent units of geometric measurement. **(S)** [**Note: The data collected could be the measures of the sides of rectangles. Or the data collected could be how many students prefer vanilla ice cream.**]
4. When creating a bar graph, a title is sufficient. **(N)** [**Note: Title, a description of each axis, and possibly a key as well as the interval units on the graph.**]
5. Addition and subtraction of fractions can be solved using the information displayed in a line plot. **(S)**
6. If the data set represents several million, the data display should be divided into at least one-million sections. **(N)** [**Note: This is the reason for scaled representations. An interval of one million being represented by 1 unit.**]

7. Line plots only show one line per graph. **(S)** [**Note: You can show comparisons.**]
8. Data displays show comparisons. **(S)**
9. Graphs are visual representations of data. **(A)**
10. When creating a visual representation for a data set, the type of graph does not matter. **(S)**

PART THREE

Planning and Implementation

CHAPTER 13

Crosswalks

The following crosswalk is included to support instructional planning. Resources can be quickly identified based upon the mathematical topic as well as type of writing and/or the strategy example given. The crosswalk identifies the mathematical topics that are included in the given examples referenced in each of the 11 writing strategies shared in the previous chapter.

DOI: 10.4324/9781003374572-16

Crosswalk of Topics and Writing Strategies

Writing Strategy	Number and Quantity	Algebraic Reasoning	Geometric Reasoning/ Measurement and Units	Data Analysis, Probability and Statistics	Universal
Always, Sometimes, and Never	X	X	X	X	
Question Quilts	X	X	X		
RAFTs	X	X	X		
Cubing/ Think Dots	X	X			
Poems	X		X		X
Journal Prompts	X	X	X		X
Writing About	X	X	X	X	
Topical Questions	X	X	X	X	
The Answer is...	X	X	X	X	
Compare/ Contrast	X	X	X	X	
Visual Prompts	X	X	X		

CHAPTER 14

Bringing It All Together

This last section provides a sample anchor task that demonstrates how several of these writing strategies can be authentically integrated into classroom instruction. A lesson plan and facilitation notes are provided.

Selling the Farm

Overview: This task was intentionally chosen because it not only models several opportunities for writing but also demonstrates the efficiency of planning for addressing multiple content and process standards. Planning for simultaneous outcomes allows time for students to dive deep into a single task rather than completing multiple tasks over the same timeframe. It allows students to make connections within the content rather than viewing the concepts as discrete and unrelated. This task is so versatile that it can be modified in unlimited ways to meet the needs of both teachers and students. The outline below is just one suggestion for how it can be facilitated.

DOI: 10.4324/9781003374572-17

153

Writing Opportunities

- ❏ Think-Write-Pair-Share (p. 5)
- ❏ Topical questions – "We're Stuck/"We're Done" (p 13)
- ❏ Problem-Solving Process (p. 7)
- ❏ Reflection (p. 13)

Content Connections: (This task can be used across grades 3–5 so possible topics may include but are not limited to)

- ❏ Fractions
- ❏ Modeling – Visual Representation
- ❏ Geometry-
    - ❏ Decomposing/composing
    - ❏ Triangles
    - ❏ Quadrilaterals
- ❏ Measurement
    - ❏ Tangram shapes
    - ❏ Area and perimeter
- ❏ Decimals
- ❏ Line/Dot-plot
- ❏ Real World Context (connections to other disciplines – land value versus cost based on area)
- ❏ Problem-solving (standards for mathematical practice)

Routines

- ❏ The five Practices for Orchestrating Productive Mathematics Discussions
- ❏ Anticipating students' solutions to a mathematics task
- ❏ Monitoring students' in-class, "real-time" work on the task
- ❏ Selecting approaches and students to share them
- ❏ Sequencing students' presentations purposefully
- ❏ Connecting students' approaches and the underlying mathematics

Materials Needed

- ❏ Tangrams (See **Preparation for Implementation: PLC Work**)
- ❏ Visual of tangram square
- ❏ Dry erase markers

- ❏ Which One Doesn't Belong Prompt
- ❏ Task handout (without embedded problem-solving process)
- ❏ Task handout (with embedded problem-solving process)
- ❏ "We're Stuck" and "We're Done" questions
- ❏ Student sample attempt #1
- ❏ Student sample attempt #2
- ❏ QUAD Reflection

Preparation

For interactive tangrams (able to be written on and large enough to manipulate) use the blackline master to make tangrams out of card stock. Laminate and cut into pieces.

Complete the task as if you are a student. *Anticipate* student strategies, solutions, misconceptions, etc. This process informed the development of the "We're Stuck" and "We're Done" questions used to support students. See **Preparation for Implementation: PLC Work** for a more detailed explanation.

Warm Up: 10 Minutes

Display the image below and ask students to determine Which One Doesn't Belong and record their reasoning. Students can do this on an index card or whiteboard. The purpose is to ensure they commit to a choice and provide rationale. Designate four corners of the room and identify each space with the letters A, B, C, or D. Ask students to go to the corner of the room that corresponds to their choice. Have students discuss their reasoning and to select a student or students to share these reasons with the rest of the class. Once all groups have shared, if there is a choice that was not selected, engage the class in determining what reasons could be used to make the case for that representation. Doing so reinforces to students that all choices are valid with reasons they do not belong. It also challenges students to look beyond the more obvious selections.

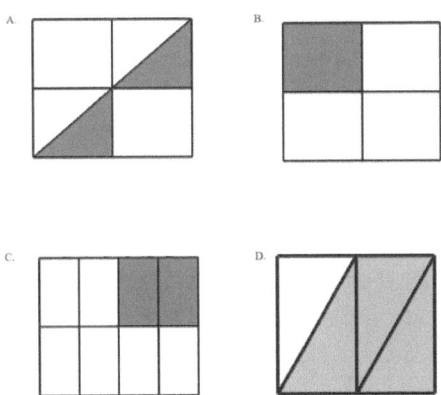

WODB

Some possible student responses may include, but are not limited to the following:

A. It is the only one partitioned into six total parts.
B. It is the only one with a singular section shaded.
C. It is the only one that divided into eight equal parts.
D. It is the only one where there is more shaded than unshaded.

Activity: 25 Minutes

Introduce the task by displaying it on the screen and reading aloud.

Distribute the handout and ask students to begin by working independently. Reinforce the expectations by displaying and stating the following:

Think about the Problem

Write by doing one of the following:

❑ Write the question that is being asked and/or create an answer statement.
❑ Write all facts you know about the problem.
❑ Write what additional information is needed or questions you have about the problem.

Providing these questions allows access for all students and eliminates "no response" as an option. NOTE: If students are using the problem-solving process graphic organizer, there is a copy of the task handout that embeds the organizer for support in the workspace section.

Have students partner to share strategies and questions. This should be done intentionally using a collaborative structure and strategy to ensure equity in each pair with both students having a voice.

Students should record any additional ideas from the discussion on the handout.

Differentiation Strategy: As students are working, allow them to access to the scaffolded questions. This can be done by having students get them from a central location (hard copy) or access on a device (electronic copy). These can also be placed on student desks while monitoring and identifying where students may need support. For example, if students choose to use the tangrams and are having difficulty making the square, ask the first question provided on the "We're Stuck" sheet, "How does making the tangram square help in working this problem?" Once they can articulate why this is necessary, provide the blackline master of the square to allow students to focus on the problem rather than putting the pieces together. As students complete the task, monitor to make sure they are utilizing the "We're Done" questions.

Also provided is a student recording sheet/chart (Tangram Chart) for working with the tangram pieces as they determine the fractional representations of each. This chart could also be used as an introductory activity for students who benefit from pre-teaching concepts.

Monitor student pairs and identify various strategies, misconceptions, "ah-ha" moments, etc. to highlight in the whole class discussion at the conclusion of the activity. *Select* (and inform) students prior to the discussion that they will be asked to share their thinking. Be specific in telling them what you would like them to share when called upon. This will keep the discussion focused on the main ideas.

Activity Synthesis: 20 Minutes

Conduct a classroom discussion by *sequencing* the student groups who were selected to share. This should be done in a manner that allows *connections* to be made between approaches and ideas to reinforce the learning goals set forth in the activity.

Student Reflection: 5 Minutes

Have students complete the QUAD Reflection independently.

- ❑ **Question:** What questions do you still have about the problem?
- ❑ **Understanding:** What do you now understand after working with this problem?
- ❑ **Activate:** How did working with a partner help you, or your partner, with this task?
- ❑ **Discourse:** What mathematical discussions/math talks were prompted by completing this task?

Preparation for Implementation: PLC Work

The best way to prepare for implementation of the task is for teachers to meet to unpack and discuss. Some ideas are outlined here.

1. Have all teachers complete the task as if they were students using the handout provided.

2. Discuss strategies and solutions from the group as well as other potential approaches. In addition, answer the questions below.

   a) Where will students demonstrate success with the task?
   b) Where will students struggle with the task? (OMG's – SREB abbreviation)
      ☐ Obstacles – Students have a lack of understanding of which strategies or procedures to apply and how those strategies work.
      ☐ Misconceptions – Students are unaware that the knowledge they have is incorrect.
      ☐ Gaps in Learning – Students lack prerequisite knowledge. (SREB, 2018)

3. Based on possible student struggles with the task, review the "stuck" questions that have been generated for support. Edit or add questions that may be needed. Keep in mind to limit the total number to between three and five so students don't get overwhelmed. Some questions you might begin with are given below.

   ❏ How does making the tangram square help in working this problem?
   ❏ What is the relationship between each of the pieces of the tangram? How do you know?
   ❏ What is the shape of the land Mr. Jones is selling? How does that help you with the task?
   ❏ What do you need to know to determine the value of each piece of land?
   ❏ What operations might you use to determine the value of each piece of land?
   ❏ What is the relationship between each of the pieces of the tangram and the whole tract of land? How will this help determine the value of each?

4. Complete the same process with the "done" questions.

   ❏ How would you justify the values you determined for each piece of land?
   ❏ How can three different geometric shapes of land have the same value?
   ❏ Why did the values of each piece of land calculate to be whole numbers?
   ❏ How would the values change if the total price were not a multiple of the denominators of the fractional pieces?

NOTE: If writing additional questions, keep in mind they...
- Arise out of students' misconceptions
- Cause the student to think more deeply about the mathematics
- Should be answerable by the teacher
- Should be answerable by more than a yes or a no
- Can be direct

5. To practice providing feedback, review Ginny's thinking about the task. Discuss the following questions (in order):

- ❑ What do you like about her work?
- ❑ What misconception(s) did she have?
- ❑ What questions could you ask her to move her thinking forward?

NOTE: The teacher provided several feedback questions. Ginny then attempted the task again. Review her second attempt and discuss the same questions from above.

Task Extensions:

- ❑ Error Analysis: Ginny's work can also be used for error analysis with students as a follow-up to the task. Ginny used multiplication of a whole number by a fraction. Younger students can use division.
- ❑ Statistics: Students create a line/dot plot ordering the fractional representations for each piece of land (tangram shape) and its relationship to the total area of the land (completed tangram square).

NOTE: Tangrams can be cut from die cuts and foam, cardstock, construction paper, etc. Alternatively, students can make their own set of tangrams through paper folding and tearing/cutting. This is a good spatial reasoning exercise. Directions are below.

Paper Folding a Tangram

- ❑ Square up a piece of 8.5″ × 11″ paper. Fold the paper so a shorter side lies on tops of (coincides) with one of the longer sides. Fold back and forth, creasing each time, and tear off the rectangle. You should now have a square piece of paper and a rectangle. Keep the square.
- ❑ Fold the square along one diagonal and crease to make two congruent right triangles. Fold back and forth, creasing each time, and tear apart the right triangles. Set one aside.

❑ Take one right triangle and fold in half so you form another set of two congruent right triangles. Fold back and forth, creasing each time, and tear apart the right triangles. Set these aside. They are the first two tangram pieces.

❑ Take the second large right triangle and position it so the right triangle is at the top and the hypotenuse is the base. Fold the right angle (the square corner) down to the middle of the opposite side (the hypotenuse). Fold back and forth, creasing each time, and tear apart the small triangle on top from the isosceles trapezoid on the bottom. Set the triangle aside. This is the third tangram piece.

❑ Turn the isosceles trapezoid so the longer base is on the bottom. Fold the left side of the trapezoid over on top of the right side so you have folded it in half along a vertical line of symmetry. Unfold and fold the left bottom corner to the middle fold line so the bottom sides lie on top of each other. Crease well and tear apart the small triangle and the remaining small square on the left of the fold line. These are the fourth and fifth pieces of the tangram.

❑ Take the remaining trapezoid and turn it so the right angles are on the left and the longer base is on the bottom. Take the upper left corner (at the obtuse angle) and fold it down and to the left corner (the lower right angle) so the bottom sides lie on top of each other. Crease well and tear apart the small triangle on the left and the remaining parallelogram. These are the sixth and seventh pieces of the tangram.

Selling the Farm
Tangram Solutions

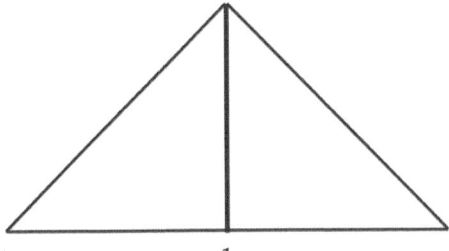

$2\ large\ triangles = \dfrac{1}{2}\ the\ tract\ of\ land$

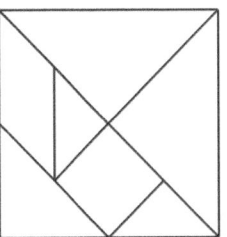

Tangram square

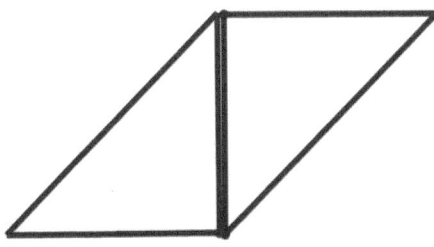

2 small triangles = 1 parallelogram
2 medium triangles = 1 large triangle

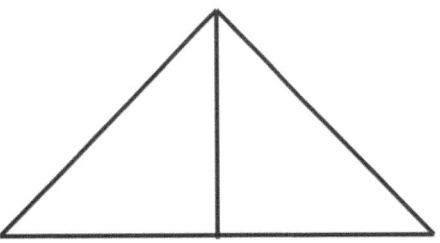

$2 \times \$525 = \$1050$

$1\ medium\ triangle = \dfrac{\$2100}{2} = \$1050$

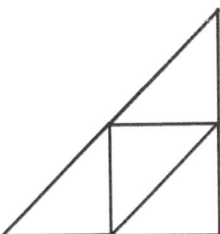

The two outside small triangles
also make a square.
4 small triangles = 1 large triangle
2 squares = 1 large triangle

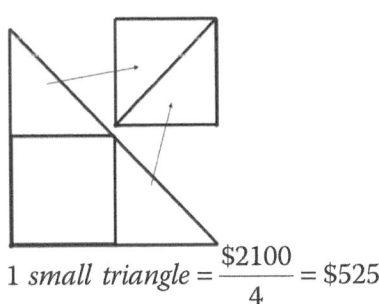

$1\ small\ triangle = \dfrac{\$2100}{4} = \$525$

$1\ square = \dfrac{\$2100}{2} = \$1050$

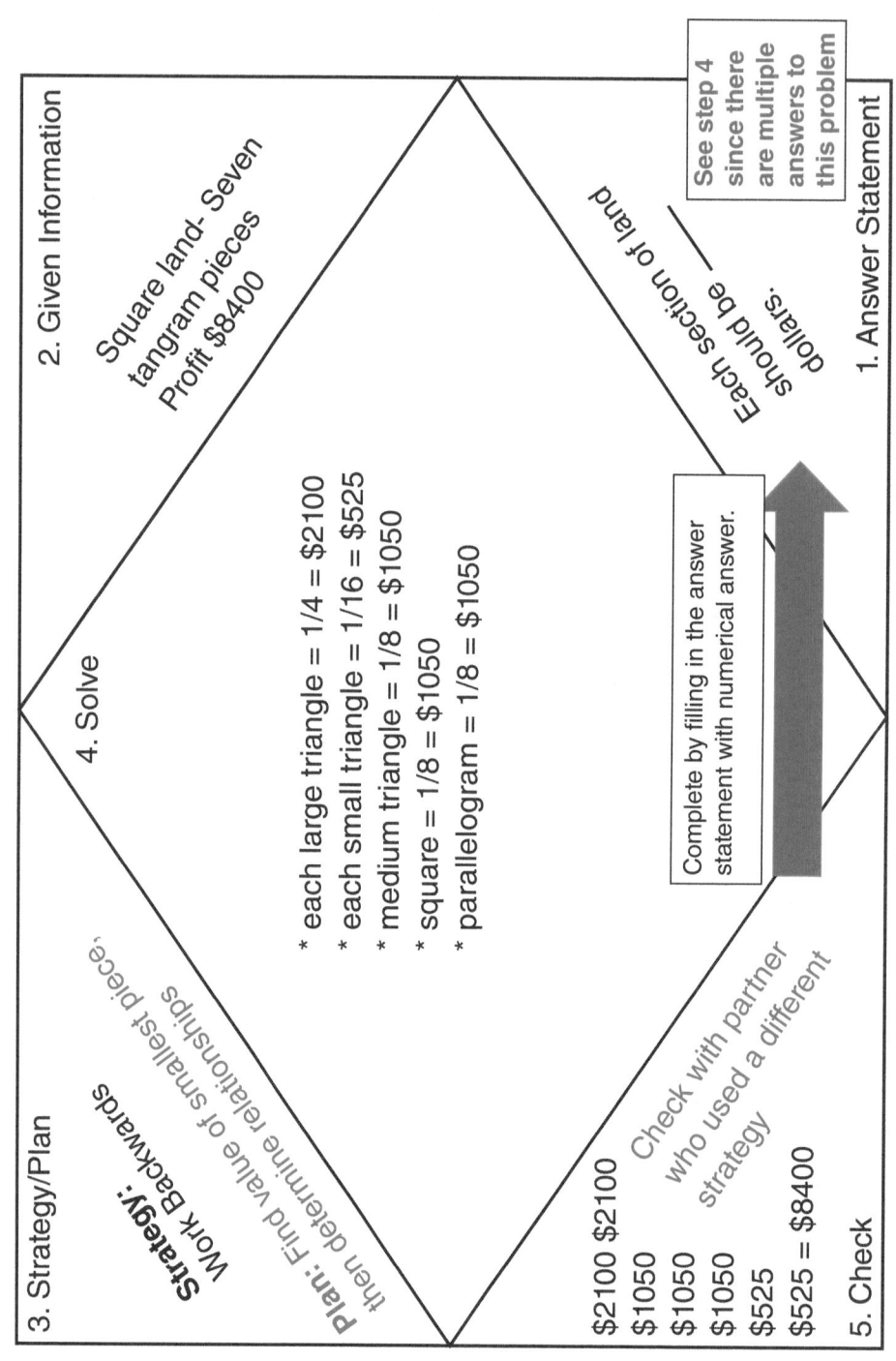

2. Given Information

Square land- Seven tangram pieces
Profit $8400

1. Answer Statement

See step 4 since there are multiple answers to this problem

Each section of land should be _____ dollars.

4. Solve

* each large triangle = 1/4 = $2100
* each small triangle = 1/16 = $525
* medium triangle = 1/8 = $1050
* square = 1/8 = $1050
* parallelogram = 1/8 = $1050

Complete by filling in the answer statement with numerical answer.

3. Strategy/Plan

Strategy: Work Backwards
Plan: Find value of smallest piece, then determine relationships.

Check with partner who used a different strategy

5. Check

$2100 $2100
$1050
$1050
$1050
$525
$525 = $8400

Tangram Master

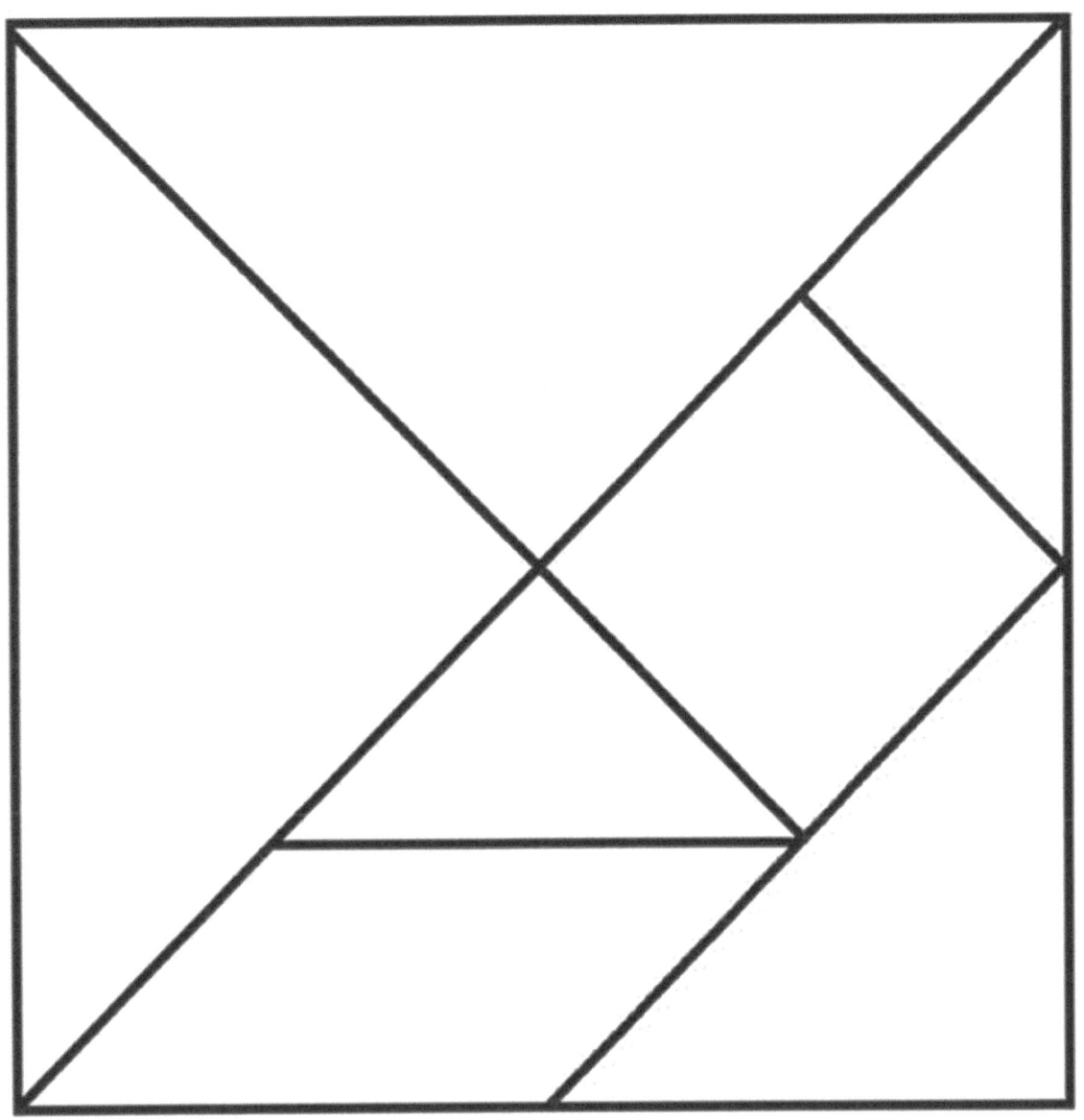

Selling the Farm
Which One Doesn't Belong

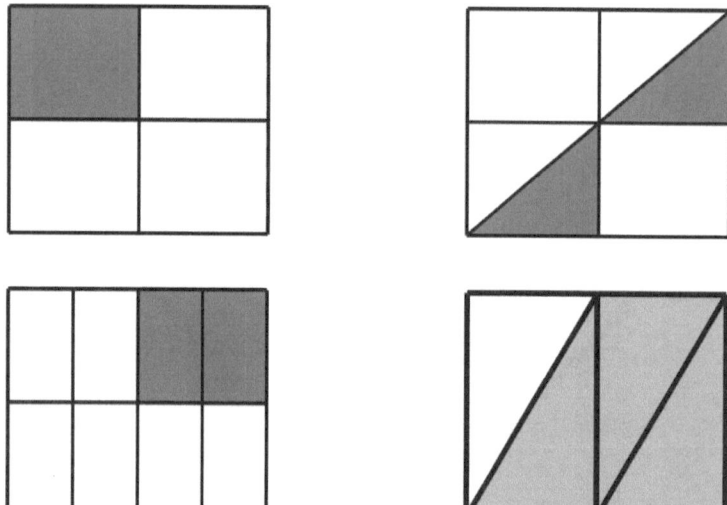

Think-Write-Pair Share

Think about the problem.

Farmer Jones is selling his land. It is the shape of a square. Mrs. Jones loves to play with tangrams. She convinced Farmer Jones to divide his land into seven pieces just like a tangram puzzle. Farmer Jones wants to make $8,400 when he sells all his land. How much should he price each section of the square?

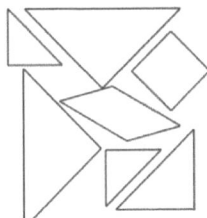

Write by doing one of the following:
If you can solve, choose a strategy, and solve.

If you cannot solve...

❑ Write the question that is being asked and/or create an answer statement
❑ Write all facts you know about the problem
❑ Write what additional information is needed or questions you have about the problem

Workspace

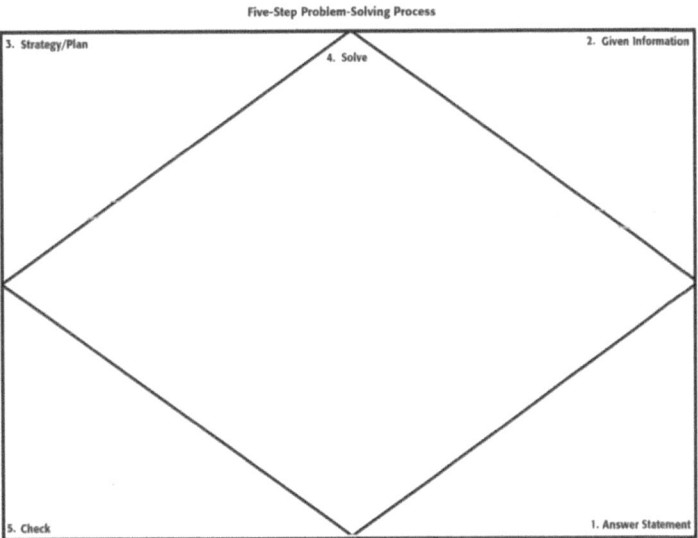

Five-Step Problem-Solving Process

3. Strategy/Plan | 4. Solve | 2. Given Information
5. Check | 1. Answer Statement

Pair with a partner and take turns discussing your strategies and solutions.

Use this space to record strategies that were different from yours.

Share various strategies & solutions with the group.

Use this space to record strategies that were different from those of you and your partner.

Selling the Farm: We're Stuck

1. How does making the tangram square help in working this problem?
2. What is the relationship between each of the pieces of the tangram? How do you know?
3. What do you need to know to determine the value of each piece of land?
4. What is the relationship between each of the pieces of the tangram and the whole tract of land? How will this help determine the value of each?

Selling the Farm: We're Stuck

1. How does making the tangram square help in working this problem?
2. What is the relationship between each of the pieces of the tangram? How do you know?
3. What do you need to know to determine the value of each piece of land?
4. What is the relationship between each of the pieces of the tangram and the whole tract of land? How will this help determine the value of each?

Selling the Farm: We're Stuck

1. How does making the tangram square help in working this problem?
2. What is the relationship between each of the pieces of the tangram? How do you know?
3. What do you need to know to determine the value of each piece of land?
4. What is the relationship between each of the pieces of the tangram and the whole tract of land? How will this help determine the value of each?

Selling the Farm: We're Stuck

1. How does making the tangram square help in working this problem?
2. What is the relationship between each of the pieces of the tangram? How do you know?
3. What do you need to know to determine the value of each piece of land?
4. What is the relationship between each of the pieces of the tangram and the whole tract of land? How will this help determine the value of each?

Selling the Farm: We're Done

1. How can three different geometric shapes of land have the same value?
2. Why did the values of each piece of land calculate to be whole numbers?
3. How would the values change if the total price were not a multiple of the denominators of the fractional pieces?

Selling the Farm: We're Done

1. How can three different geometric shapes of land have the same value?
2. Why did the values of each piece of land calculate to be whole numbers?
3. How would the values change if the total price were not a multiple of the denominators of the fractional pieces?

Selling the Farm: We're Done

1. How can three different geometric shapes of land have the same value?
2. Why did the values of each piece of land calculate to be whole numbers?
3. How would the values change if the total price were not a multiple of the denominators of the fractional pieces?

Selling the Farm: We're Done

1. How can three different geometric shapes of land have the same value?
2. Why did the values of each piece of land calculate to be whole numbers?
3. How would the values change if the total price were not a multiple of the denominators of the fractional pieces?

Tangram Chart: Use the tangrams to complete the following.

Tangram Piece	Fraction of the Whole
Large triangle	
Small triangle	
Medium triangle	
Parallelogram	
Square	

Use the fractional values above to complete the chart.

Tangram Pieces	Expression	Fraction of the Whole Seven-Piece Square
1. Both large triangles		
2. Small triangle and square		
3. Medium triangle and square		
4. Large triangle minus medium triangle		
5. Parallelogram minus small triangle		
6. Four large triangles		
7. Twelve small triangles		
8. Half of a large triangle		
9. One fourth of a large triangle		
10. All seven pieces		

On the back, group the pieces according to their area from smallest to largest. Then trace each piece in its group.

Student Sample Attempt #1

Five-Step Problem-Solving Process

2. Given Information
- divided into seven pieces
- $8,400

1. Answer Statement

Farmer Jones

Should price each section of the square $12.00.

Each piece is $\frac{1}{7}$ of the whole.

4. Solve

$$7\overline{)8,400} = 1,200$$

-7
14
-14
000
-0
00
-0
0

3. Strategy/Plan

My plan is to divide.

5. Check

$$\begin{array}{r} 1,200 \\ \times \quad 7 \\ \hline 8,400 \end{array}$$

Student Sample Attempt #2

Five-Step Problem-Solving Process

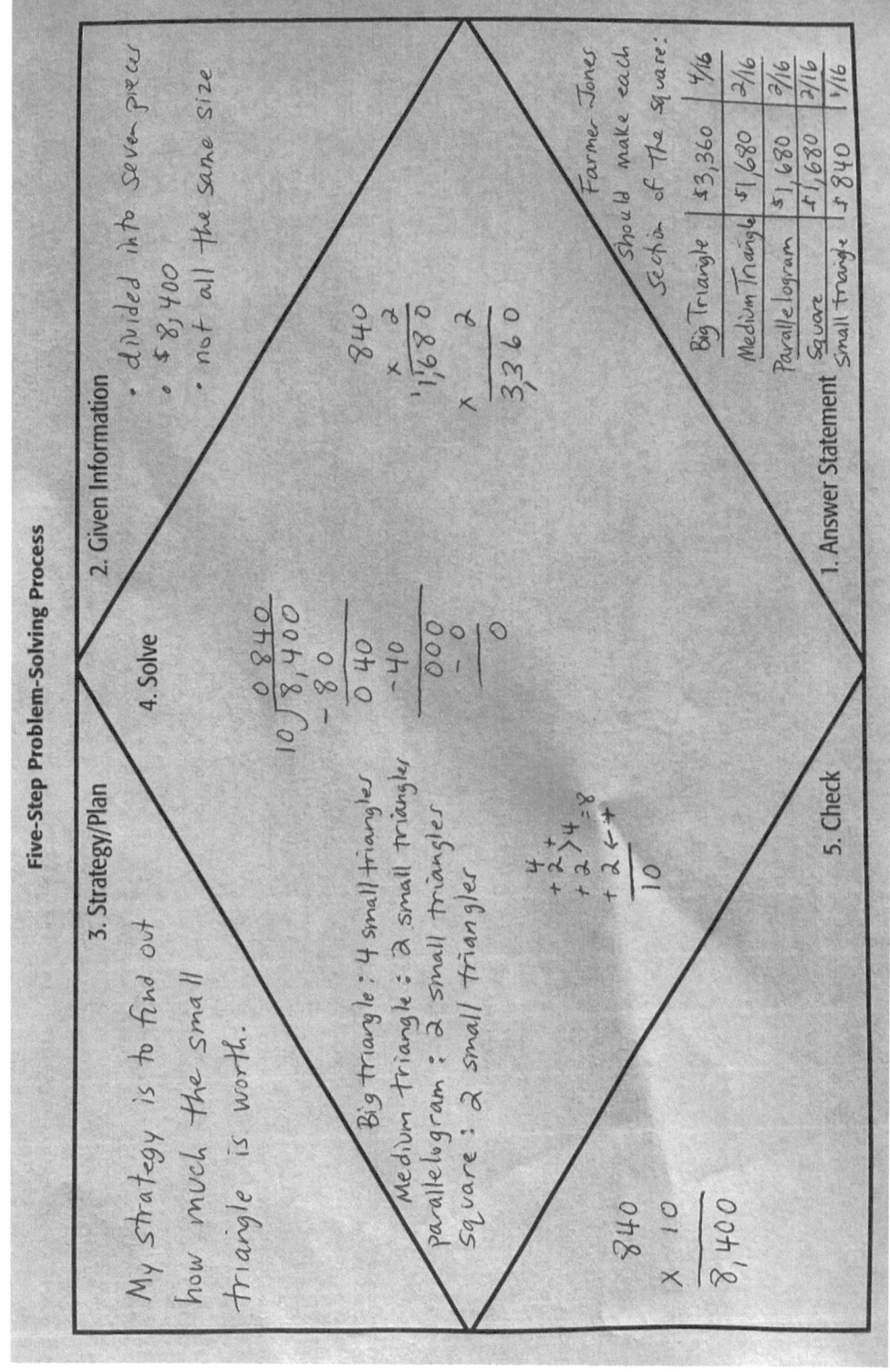

3. Strategy/Plan

My strategy is to find out how much the small triangle is worth.

Big triangle : 4 small triangles
Medium triangle : 2 small triangles
Parallelogram : 2 small triangles
Square : 2 small triangles

2. Given Information

- divided into seven pieces
- $8,400
- not all the same size

4. Solve

$$840$$
$$\times\ 2$$
$$\overline{1,680}$$
$$\times\ 2$$
$$\overline{3,360}$$

$$\begin{array}{r}0\ 840\\10\overline{)8,400}\\-8\ 0\\\hline 0\ 40\\-40\\\hline 000\\-0\\\hline 0\end{array}$$

$$\begin{array}{r}4\\+2\\+2\\+2\\\hline 10\end{array}$$

4 → 4 small triangles

+2 →4=8

5. Check

$$840$$
$$\times\ 10$$
$$\overline{8,400}$$

1. Answer Statement

Farmer Jones should make each section of the square:

Big Triangle	$3,360	4/16
Medium Triangle	$1,680	2/16
Parallelogram	$1,680	2/16
Square	$1,680	2/16
Small Triangle	$840	1/16

QUAD Reflection

Q: Question:
What questions do you still have about the task?

U: Understanding:
What do you now understand after working with this task?

A: Activate:
How did working with a partner help you, or your partner, with this task?

D: Discourse:
What mathematical discussions/math talks were prompted by completing this task?

QUAD Reflection

Q: Question:
What questions do you still have about the task?

U: Understanding:
What do you now understand after working with this task?

A: Activate:
How did working with a partner help you, or your partner, with this task?

D: Discourse:
What mathematical discussions/math talks were prompted by completing this task?

Afterword

We have always had educators ask us if they could get a list of the questions we used during the day's training, a list of the writing prompts we used, an example of a DI strategy we mentioned, etc. Our questions came from our interactions with the participants; the writing prompts, while mostly intentional, also might have been inspired by a comment or question from a participant, and, yes, we did have examples of the various strategies that we often mention. At the same time, our editor, Lauren, to whom we owe so much thanks and gratitude, asked us if we would ever consider doing a book of consumables for educators to use. Hence, the seed was sown that finally grew into what became this book series.

Everyone acknowledges that communicating in mathematics is essential. Communication was one of the original five process standards of The National Council of Teachers of Mathematics. Over the years, we have collected, found, and created many classroom resources that provide authentic opportunities for students to communicate mathematically. So, the next step was culling through the many resources we had used and developed over the years. We sorted, resorted, looked at, accepted some, rejected others, and even created new ones as needed. We wanted to offer this series in four books to meet the individual needs of the various grade bands. At the same time, we wanted to provide examples and prompts that would cover the breadth of the grade band's mathematical topics and provide materials to support deepening the understanding of the topics.

Afterword

Then to the writing and pulling together of the resources, which for us as mathematicians, the latter was much less challenging. This is ironic since this book focuses on writing and communicating in mathematics! What emerged is what you have on these pages. We hope that this resource becomes a go-to to meet your everyday classroom needs for providing opportunities for your students to engage in communicating about mathematics and communicating mathematically. Take what we have provided, expand on it, and make it your own. As you reimagine, retool, and even create your versions, we ask that you reach out and share. We would love to hear from you. You can find us at https://tljconsultinggroup.com/about-us/tammy-jones/ and https://leslietexasconsulting.com/.

Bibliography

Growney, J. *Intersections—Poetry with Mathematics.* https://poetrywithmathematics. blogspot.com

Jones, T. L. and Texas, L. A. (2017). *Strategic Journeys for Building Logical Reasoning: Activities across the Content Areas.* New York: Routledge, Taylor & Francis Group.

Smith, M. S. and Stein, M. K. (2018). *5 Practices for Orchestrating Productive Mathematics Discussions.* Reston, VA: National Council of Teachers of Mathematics, Inc.

Southern Regional Education Board. (2018). *Making Math Matter: High-Quality Assignments That Help Students Solve Problems and Own Their Learning.* SREB. https://www.sreb.org/sites/main/files/file-attachments/18v04_math_matters_report_final.pdf?1521473373

Texas, L. A. and Jones, T. L. (2013). *Strategies for Common Core Mathematics: Implementing the Standards for Mathematical Practice.* New York: Routledge, Taylor & Francis Group.